MONOGRAPHS ON STATISTICS AND APPLIED PROBABILITY

General Editors

D.R. Cox, D.V. Hinkley, N. Reid, D.B. Rubin and B.W. Silverman

1 Stochastic Population Models in Ecology and Epidemiology *M.S. Bartlett* (1960)

2 Queues *D.R. Cox and W.L. Smith* (1961)

3 Monte Carlo Methods *J.M. Hammersley and D.C. Handscomb* (1964)

4 The Statistical Analysis of Series of Events *D.R. Cox and P.A.W. Lewis* (1966)

5 Population Genetics *W.J. Ewens* (1969)

6 Probability, Statistics and Time *M.S. Bartlett* (1975)

7 Statistical Inference *S.D. Silvey* (1975)

8 The Analysis of Contingency Tables *B.S. Everitt* (1977)

9 Multivariate Analysis in Behavioural Research *A.E. Maxwell* (1977)

10 Stochastic Abundance Models *S. Engen* (1978)

11 Some Basic Theory for Statistical Inference *E.J.G. Pitman* (1978)

12 Point Processes *D.R. Cox and V. Isham* (1980)

13 Identification of Outliers *D.M. Hawkins* (1980)

14 Optimal Design *S.D. Silvey* (1980)

15 Finite Mixture Distributions *B.S. Everitt and D.J. Hand* (1981)

16 Classification *A.D. Gordon* (1981)

17 Distribution-free Statistical Methods *J.S. Maritz* (1981)

18 Residuals and Influence in Regression *R.D. Cook and S. Weisberg* (1982)

19 Applications of Queueing Theory *G.F. Newell* (1982)

20 Risk Theory, 3rd edition *R.E. Beard, T. Pentikainen and E. Pesonen* (1984)

21 Analysis of Survival Data *D.R. Cox and D. Oakes* (1984)

22 An Introduction to Latent Variable Models *B.S. Everitt* (1984)

23 Bandit Problems *D.A. Berry and B. Fristedt* (1985)

24 Stochastic Modelling and Control *M.H.A. Davis and R. Vinter* (1985)

25 The Statistical Analysis of Compositional Data *J. Aitchison* (1986)

26 Density Estimation for Statistical and Data Analysis *B.W. Silverman* (1986)

27 Regression Analysis with Applications *G.B. Wetherill* (1986)

28 Sequential Methods in Statistics, 3rd edition *G.B. Wetherill* (1986)

29 Tensor Methods in Statistics *P. McCullagh* (1987)

30 Transformation and Weighting in Regression *R.J. Carroll and D. Ruppert* (1988)

31 Asymptotic Techniques for use in Statistics *O.E. Barndorff-Nielsen and D.R. Cox* (1989)

32 Analysis of Binary Data, 2nd edition *D.R. Cox and E.J. Snell* (1989)

33 Analysis of Infectious Disease Data *N.G. Becker* (1989)

34 Design and Analysis of Cross-Over Trials *B. Jones and M.G. Kenward* (1989)

35 Empirical Bayes Method, 2nd edition *J.S. Maritz and T. Lwin* (1989)

36 Symmetric Multivariate and Related Distributions *K.-T. Fang, S. Kotz and K. Ng* (1989)

37 Generalized Linear Models, 2nd edition *P. McCullagh and J.A. Nelder* (1989)

38 Cyclic Designs *J.A. John* (1987)

39 Analog Estimation Methods in Econometrics *C.F. Manski* (1988)

40 Subset Selection in Regression *A.J. Miller* (1990)

41 Analysis of Repeated Measures *M. Crowder and D.J. Hand* (1990)

42 Statistical Reasoning with Imprecise Probabilities *P. Walley* (1990)

43 Generalized Additive Models *T.J. Hastie and R.J. Tibshirani* (1990)

44 Inspection Errors for Attributes in Quality Control *N.L. Johnson, S. Kotz and X. Wu* (1991)

45 The Analysis of Contingency Tables, 2nd edition *B.S. Everitt* (1992)

46 The Analysis of Quantal Response Data *B.J.T. Morgan* (1992)

47 Longitudinal Data with Serial Correlation: A State-space Approach *R.H. Jones* (1993)

48 Differential Geometry and Statistics *M.K. Murray and J.W. Rice* (1993)

49 Markov Models and Optimization *M.H.A. Davis* (1993)

(Full details concerning this series are available from the Publishers.)

Differential Geometry and Statistics

Michael K. Murray

Department of Pure Mathematics
The University of Adelaide
Adelaide
Australia

and

John W. Rice

School of Information Science and Technology
The Flinders University of South Australia
Bedford Park
Australia

CHAPMAN & HALL

London · Glasgow · New York · Tokyo · Melbourne · Madras

Published by Chapman & Hall, 2–6 Boundary Row, London SE1 8HN

Chapman & Hall, 2–6 Boundary Row, London SE1 8HN, UK

Blackie Academic & Professional, Wester Cleddens Road, Bishopbriggs, Glasgow G64 2NZ, UK

Chapman & Hall Inc., 29 West 35th Street, New York NY10001, USA

Chapman & Hall Japan, Thomson Publishing Japan, Hirakawacho Nemoto Building, 6F, 1–7–11 Hirakawa-cho, Chiyoda-ku, Tokyo 102, Japan

Chapman & Hall Australia, Thomas Nelson Australia, 102 Dodds Street, South Melbourne, Victoria 3205, Australia

Chapman & Hall India, R. Seshadri, 32 Second Main Road, CIT East, Madras 600 035, India

First edition 1993

Printed in Great Britain

ISBN 0 412 39860 5

A catalogue record for this book is available from the British Library

Library of Congress Cataloging-in-Publication data available

Printed on permanent acid-free text paper, manufactured in accordance with the proposed ANSI/NISO Z 39.48-199X and ANSI Z 39.48-1984

For

Patty and Sam,

and

Loretta, Benno and Thomas

Contents

Preface xi

1. The geometry of exponential families **1**
1.1 Geometry, parameters and co-ordinates 1
1.2 Canonical co-ordinates 4
1.3 Affine spaces 6
1.4 Log-likelihood and affine structure 9
1.5 When is a family exponential? 14
1.5.1 A geometric criterion 14
1.5.2 A computable criterion 18
1.6 Parameter independence 20
1.7 Remarks for Chapter 1 23
1.8 Exercises for Chapter 1 24

2. Calculus on manifolds **25**
2.1 Introduction 25
2.2 The basic apparatus 26
2.2.1 Functions, variables and parameters 26
2.2.2 Rates of change under variations 29
2.2.3 The chain rule, velocities and tangent vectors 35
2.2.4 Co-ordinate independence 42
2.2.5 Differentials and 1-forms 45
2.2.6 Vector fields 48
2.2.7 Co-ordinate calculations 50
2.2.8 Differentiability and the chain rule 54
2.2.9 Manifolds 57
2.2.10 Co-ordinates and geometry 59
2.3 Remarks for Chapter 2 62
2.4 Exercises for Chapter 2 62

3. Statistical manifolds **63**
3.1 Realising manifolds 63
3.1.1 Smooth maps 64
3.1.2 The derivative map 67
3.1.3 Submanifolds 68
3.1.4 Inclusions, immersions and embeddings 72
3.2 The definition of a statistical manifold 76
3.2.1 Statistical manifolds and the score 76
3.2.2 Some useful formulae 79
3.2.3 The second fundamental form of a family 80
3.3 Curved exponential families 83
3.4 Lie groups 84
3.4.1 Lie group actions 87
3.4.2 Transformation models 91
3.5 Remarks for Chapter 3 95
3.6 Exercises for Chapter 3 96

4. Connections **97**
4.1 Introduction: connections and geometry 97
4.2 Rates of change of vector fields on the plane 100
4.3 Affine spaces and flat connections 109
4.4 Connections on submanifolds of affine spaces 111
4.4.1 The second fundamental form 116
4.5 Amari's 1-connection 117
4.5.1 The second fundamental form 119
4.6 Amari's α-connection 120
4.7 Connections on the cotangent bundle 121
4.8 Dual connections and symmetry 124
4.9 Geodesics and the exponential map 126
4.9.1 The second fundamental form and geodesics 130
4.10 Remarks for Chapter 4 131
4.11 Exercises for Chapter 4 131

5. Curvature **132**
5.1 Introduction 132
5.1.1 Parallel translation 135
5.1.2 Differential 2-forms 139
5.1.3 Curvature 148
5.1.4 Vanishing curvature and flatness 154
5.2 Remarks for Chapter 5 155

5.3 Exercises for Chapter 5 155

6. Information metrics and statistical divergences 157

6.1 Introduction 157
6.2 Riemannian metrics 161
6.3 Metric preserving connections 165
6.4 Dual connections 170
6.5 Statistical divergences 173
6.5.1 Orthogonality and divergence minimisation 176
6.6 Transformation models 179
6.7 Other geometries 181
6.7.1 The observed Fisher information metric 181
6.7.2 Yokes 184
6.7.3 Preferred point geometry 185
6.8 The square root likelihood 185
6.8.1 The Levi-Civita connection and the square root likelihood 187
6.9 Integration and densities on a manifold 188
6.9.1 Half-densities and the square root likelihood 191
6.10 Remarks for Chapter 6 192
6.11 Exercises for Chapter 6 192

7. Asymptotics 194

7.1 Asymptotics 194
7.1.1 Introduction 194
7.2 Estimators 195
7.2.1 Introduction: unbiased estimators 195
7.3 The maximum likelihood estimator 196
7.3.1 The mle of an exponential family 197
7.3.2 The mixture affine structure 200
7.4 The Cramer–Rao inequality 201
7.5 Sufficiency 204
7.5.1 Conditional distributions 204
7.5.2 Sufficient statistics: the factorisation principal 206
7.5.3 Minimal sufficiency 208
7.5.4 The likelihood ratio statistic 208
7.6 The central limit theorem and the score form 211
7.7 Asymptotics of the maximum likelihood estimator 213
7.7.1 The asymptotics of the unnormalised mle 213
7.7.2 The asymptotics of the mle 213

7.8 Asymptotics and exponential families 216
7.8.1 Curved exponential families 219
7.9 Barndorff-Nielsen's p^* formula 221
7.10 Remarks for Chapter 7 222
7.11 Exercises for Chapter 7 222

8. Bundles and tensors **223**
8.1 Introduction 223
8.2 Tangent bundles and vector bundles 223
8.3 New vector bundles from old 229
8.4 Connections on a vector bundle 230
8.5 Frame bundles and principal bundles 231
8.6 Associated bundles 235
8.7 Tensors 239
8.8 Remarks for Chapter 8 241
8.9 Exercises for Chapter 8 242

9. Higher order geometry **243**
9.1 Introduction 243
9.2 Jets and jet bundles 243
9.3 Taylor series and co-ordinate strings 248
9.3.1 Taylor series 248
9.3.2 Co-ordinate strings 250
9.3.3 Components of a co-ordinate string 252
9.3.4 Flatness of coordinate strings 252
9.3.5 Co-ordinate strings and the infinite frame bundle 254
9.4 Strings and phylon representations 254
9.4.1 Phylon representations 255
9.5 Remarks for Chapter 9 262
9.6 Exercises for Chapter 9 263

References **264**

Notation index **267**

Subject index **270**

Preface

Several years ago our statistical friends and relations introduced us to the work of Amari and Barndorff-Nielsen on applications of differential geometry to statistics. This book has arisen because we believe that there is a deep relationship between statistics and differential geometry and moreoever that this relationship uses parts of differential geometry, particularly its 'higher-order' aspects not readily accessible to a statistical audience from the existing literature. It is, in part, a long reply to the frequent requests we have had for references on differential geometry! While we have not gone beyond the path-breaking work of Amari and Barndorff-Nielsen in the realm of applications, our book gives some new explanations of their ideas from a first principles point of view as far as geometry is concerned. In particular it seeks to explain why geometry should enter into parametric statistics, and how the theory of asymptotic expansions involves a form of higher-order differential geometry.

The first chapter of the book explores exponential families as flat geometries. Indeed the whole notion of using log-likelihoods amounts to exploiting a particular form of flat space known as an affine geometry, in which straight lines and planes make sense, but lengths and angles are absent. We use these geometric ideas to introduce the notion of the second fundamental form of a family whose vanishing characterises precisely the exponential families.

The second chapter, in which we introduce manifolds, should be most useful to statisticians who want to learn about the subject. The traditional theory starts with a heavy meal of the purest mathematics, (topological spaces, co-ordinate coverings, differentiable functions), before embarking on a treatment of calculus that is filled with multilinear algebra, and bears little relationship to anything one might have learned about several-variable calculus as an undergraduate. By contrast our treatment starts with calculus on manifolds as a geometrical approach to the theory of rates of change of functions, treating it as though it were a first course on several variable calculus. We explain how the several-variable chain rule can be interpreted as dividing variations through a point into families with different velocities, how *df* is to be interpreted as the rate of change of *f* as a function of velocity, and what are vector fields (contravariant 1-tensors) and 1-forms (covariant 1-tensors). We give a brief discussion of the foundational concepts of differentiability and manifolds at the end of the chapter, but these are not really important for the application of differential geometry to statistics.

Our comment on the great divide between the so-called co-ordinate-free and index-laden approaches to differential geometry, is that we aim to be geometrical without being obsessed with freedom from co-ordinates. We have enormous interest in co-ordinates when it comes to calculations. However, it seems pointless to us to be in the position either to be able to calculate everything but explain nothing, or to explain everything but calculate nothing. So we explain geometrical concepts in co-ordinate-free terms, and we translate them into co-ordinate systems for calculations, with whatever debauches of indices they require.

Once the basic notions are in place, most notably the definition in Chapter 2 of the tangent space to a manifold, we begin an elaboration of the parts of differential geometry that are useful in statistics, illustrating them with statistical applications and examples. As the number of statistical applications is growing rapidly we have been unable to consider them all. However we believe that we have covered all the *concepts* from differential geometry that are needed at this point in time. Chapter 3 explains the idea of submanifold and the definition of a statistical manifold. We mention again the simplest statistical manifolds, the exponential families, and then consider the families with a high degree of symmetry, the transformation models.

The next two chapters introduce the concept of connections and their curvature, Amari's α-connections and the theory of statistical divergences. A connection defines the rate of change of vector fields. It therefore tells us which curves have constant tangent vector fields, that is which curves are straight lines or geodesics. Hence a connection defines a notion of geometry, or straight lines and the different connections define different geometries. Some connections are essentially 'flat'. That is, the geometry they define is Euclidean. The curvature of a connection is a measure of its departure from flatness.

In Chapter 6 we consider the theory of Riemannian manifolds. An initial impetus for introducing differential geometry into statistics was the observation of Rao that the Fisher information could be interpreted as a Riemannian metric on the space of parametrised probability distributions forming the statistical model.

Chapter 7 introduces the maximum likelihood estimator and considers some results in asymptotics, in particular the work of Amari. Here we begin to see the importance of Taylor series and the need for a higher-order geometry in statistics. The final Chapters 8 and 9 consider this higher order geometry: the theory of strings or phyla developed by Barndorff-Neilsen and Blæsild. Strings are generalisations of tensors. If we think of tensors in co-ordinates as functions with many indices transforming under a change of co-ordinates by the first derivative of the co-ordinate transformation, then a string has more indices and transforms by higher derivatives of the co-ordinate transformation. To consider strings from a co-ordinate-free point of view requires that we introduce in Chapter 8 the theory of principal and vector bundles, in particular the so-called infinite frame bundle and the infinite phylon group. Chapter 9 then applies this theory to Taylor expansions and co-ordinate strings and relates the theory of strings to the representation theory of the infinite phylon group.

A book is not just the result of the labours of its authors but also of the generosity of others. First and foremost we thank our families who had to live through this book's production; then our many statistical colleagues who have laboured to explain their subject to us. Our thanks and apologies for the places where despite your efforts we get it wrong. Particular thanks must go to Peter McCullagh for providing us with TEX macros for this book and to Peter Jupp for his amazingly thorough reading of our first manuscript. Of course any remaining errors and omissions are our responsibility.

Michael K. Murray
John W. Rice

CHAPTER 1

The geometry of exponential families

1.1 Geometry, parameters and co-ordinates

Parametric statistics concerns parametrised families of probability distributions $p(\theta)$, where the parameter $\theta = (\theta^1, \ldots, \theta^d)$ varies over some open set in $\mathbf{R}^d$. The most common example is the normal family, which is usually expressed as a family of densities

$$p(\mu, \sigma) = N(\mu, \sigma^2) = \frac{1}{\sqrt{2\pi}\sigma} \exp(\frac{-(x-\mu)^2}{2\sigma^2})$$

The parameter θ in this case is the pair (μ, σ) which varies over the open subset of $\mathbf{R}^2$ determined by $\mu > 0$. The sample space is $\mathbf{R}$ and the densities are with respect to Lebesgue measure dx on $\mathbf{R}$, so that as a set of probability measures the normal family is

$$\mathcal{N} = \{p(\mu, \sigma)dx \mid \mu \in \mathbf{R}, \sigma > 0\}$$

Statistical inference concerns situations in which one knows or suspects that data are generated by sampling from a space according to a probability distribution which is a member of some known family $p(\theta)$. The problem is to infer facts about the distribution from the data. For example, one might want to know the parameter value of the distribution (point estimation), or simply whether or not this value lies in some particular set of parameters (hypothesis testing). If a given collection of numbers has arisen by sampling from a normal distribution then one might ask which normal distribution it is, i.e. what are the values of μ and σ for this particular normal distribution. On the other hand one might only want to test the hypothesis that the mean is greater than 1.

Many of these tests and much of the theory of statistical inference depends on the choice of parameters. This dependence on parameters usually comes about because the theory applies differential calculus to these parameters; differentiating them or perhaps Taylor series expanding some function with respect to them. It is important to know how the theory depends on the parameters, either because one suspects that it should not depend on the parameters at all or because one would like to know if a particular choice of parameters may simplify matters. That part of differential geometry which we have called 'Calculus on Manifolds' in Chapter 2, is concerned exactly with this question of how the differential calculus depends on co-ordinates. It obviously has immediate application to these problems.

Differential geometry however is more that just understanding how calculus depends on co-ordinates – it is also a theory of geometry or shape. Borrowing from the ideas of differential geometry we think of families of probability distributions as entities independent of any particular parametrisation, and able to support a variety of geometries. We seek to relate their statistical properties to these geometries.

We will motivate this use of geometry in statistics by considering in this first chapter the geometric significance of the exponential family. It has long been known that in seeking answers to many statistical questions the simplest type of family to deal with is the *exponential family*, i.e. one which can be parametrised in the form

$$p(\theta) = \exp\big(\theta^1 x^1 + \ldots + \theta^r x^r - K(\theta)\big)\, d\mu$$

where $x^1, \ldots, x^n$ are random variables and μ a measure on some sample space. If there is to be a meaningful relationship between statistics and geometry we must be able to discover some geometric significance to a family being exponential. Indeed we can. We shall show that the geometry of an exponential family is perhaps the simplest geometry possible, that is, affine geometry. This explains the geometric significance of the so-called *canonical parameters* θ in the exponential parametrisation. They are the affine co-ordinates arising from the affine geometry.

If we regard a parametrised family of probability distributions as analogous to a surface with a co-ordinate system on it then individual probability distributions correspond to the points on the surface, and their parameter values are their co-ordinates.

After a fashion, this kind of interpretation is often made for the normal family by regarding (μ, σ) as the Cartesian co-ordinates of a point in the upper half plane in $\mathbf{R}^2$. This certainly sets up a correspondence between the probability distributions of the normal family and points on a surface, viz. the upper half plane. However, there is no good reason to think that such an *ad hoc* correspondence should be taken seriously, nor in particular that the flatness of the plane or any other of its geometric features should have any significance for the statistical properties of the normal distribution. On the other hand, as we shall see in this first chapter, if we use the exponential parametrisation to set up a correspondence between individual normal distributions and points of a plane then we should take its flatness very seriously indeed.

Ultimately we shall be considering a variety of geometries on any given family of probability distributions, which may be interpreted loosely as imposing a variety of shapes upon the 'surface' of probability distributions. For any given geometry, or shape, there may be co-ordinate systems which are closely tied to the geometry and co-ordinate systems which are not. For example, the Euclidean geometry of a plane, involving such concepts as distance and orthogonality, is well reflected by Cartesian co-ordinate systems, but not so well by polar or other kinds. The existence of an exponential parametrisation comes about because, in a certain well defined sense, the family of probability distributions is flat, and the exponential parameters or co-ordinates are the ones adapted to this flatness.

It is well known that a family can be exponential without it being immediately apparent. For example, the normal family is an exponential family since as well as its usual parametrisation it can be parametrised

$$p(\theta^1, \theta^2)(x) = \exp(x^2\theta^1 + x\theta^2 - K(\theta))$$

where

$$\theta^1 = \frac{-1}{2\sigma^2}, \quad \theta^2 = \frac{\mu}{\sigma^2} \quad \text{and} \quad K(\theta) = \frac{1}{2}\log(\frac{-\pi}{\theta^1}) - \frac{(\theta^2)^2}{4\theta^1}$$

The parameter $\theta = (\theta^1, \theta^2)$ is called the canonical parameter, and lies in the open subset of $\mathbf{R}^2$ defined by $\theta^1 < 0$.

The point is that one cannot say that a family of probability distributions is not an exponential family just because it does not appear in exponential form. It has to be proved that the family cannot be reparametrised into exponential form. The question as to whether or not a family of probability distributions is an exponential family is therefore a question about reparametrisation of the family.

Because we are able to assign geometric meaning to a family being exponential we are able to produce an invariant of any family, its second fundamental form, whose vanishing characterises exactly the exponential families. In the case of a one-dimensional family this second fundamental form is closely related to Efron's statistical curvature Efron (1975).

1.2 Canonical co-ordinates

Let us begin to look for geometry in an exponential family

$$p(x, \theta) = \exp(\sum_{i=1}^{r} x^i \theta^i - K(\theta)) \tag{1.2.1}$$

by considering the canonical parameters $\theta = (\theta^1, \ldots, \theta^r) \in \mathbf{R}^r$ which are obviously not arbitrary but have to be chosen so that p has this special form. We seek to relate the fact that an exponential family has this restricted set of parameters to some kind of geometry. It is important then to know how restricted this set of parameters is. That is 'how canonical are the canonical parameters ?' Is it possible to have another set of random variables $y^i(x)$, parameters $\phi^i(\theta)$, and a function $J(\theta)$ such that

$$p(x, \theta) = \exp(\sum_{i=1}^{r} y^i(x)\phi^i(\theta) - J(\theta))?$$

We should, of course, also allow the possibility that we have changed the measure relative to which these are densities, so we could have

$$p(x, \theta) = \exp(\sum_{i=1}^{r} y^i(x)\phi^i(\theta) - J(\theta) + f(x)) \tag{1.2.2}$$

for some function f. Comparing (1.2.1) and (1.2.2) we see that we must have

$$\sum_{i=1}^{r} x^i\theta^i - K(\theta) = \sum_{i=1}^{r} y^i(x)\phi^i(\theta) - J(\theta) + f(x) \tag{1.2.3}$$

and differentiating both sides of (1.2.3) with respect to x^i gives

$$\theta^i = \sum_{j=1}^{r} \frac{\partial y^j}{\partial x^i}\phi^j + \frac{\partial f}{\partial x^i}$$

for every $i = 1, \ldots, r$. In particular if we choose a point θ such that $\phi^i(\theta) = 0$ for all i we see that

$$\xi^i = \frac{\partial f}{\partial x^i}$$

must be a constant vector and

$$X_i^j = \frac{\partial y^j}{\partial x^i}$$

must be a constant matrix.

The two sets of canonical parameters are therefore related by

$$\theta^i = \sum_{j=1}^{r} X_j^i\phi^j + \xi^i \tag{1.2.4}$$

The relationship in equation (1.2.4) between the canonical parameters of two exponential parametrisations is exactly the same as the relationship between Cartesian co-ordinate systems in the plane, but with the restriction in this latter case that the 2×2 matrix X_j^i must be a rotation matrix. In order to obtain general non-singular matrices X_j^i we must go beyond Cartesian systems to those determined by skewed axes with independent units of length. Such co-ordinate systems are not tied to the notions of length and angle, but they do reflect the notions of straightness and parallelism, or equivalently, as we shall explain, the notion of parallel translation in the plane.

As an example of the kind of geometry implied by this relationship between the canonical co-ordinates notice that we can use the canical co-ordinates to define the notion of a straight line in an exponential family. We say that a subset L of an exponential family is a line if its image under some canonical co-ordinates is a line in $\mathbf{R}^r$. This is, in fact, independent of which particular canonical co-ordinates are chosen because the image of a line under an affine transformation is still a line. Similarly we can define an affine subspace of an exponential family to be a subset whose image under some (and hence all) canonical co-ordinates is an affine subspace of $\mathbf{R}^r$, that is, the translate of a vector subspace. To understand where all this geometry is coming from we have to introduce the concept of an *affine space.*

1.3 Affine spaces

An affine space can be thought of as a set which becomes a vector space by selecting a point to be the zero point. The plane is an important motivating example. It is not a vector space itself, and in particular no one point stands out as the zero element. However, having chosen an arbitrary point to play the role of an origin, and so to be the zero vector, all of the other points can be regarded as vectors. Specifically, points correspond to the tips of arrows based at the chosen origin, and they are added or multiplied by scalars according to the parallelogram rules applied to their corresponding arrows. Although it usually doesn't matter, the addition and scalar multiplication of points is completely different for different choices of origin, because of the different arrows to which points correspond.

A vector in the plane doesn't usually refer to a single arrow but rather to a whole family of arrows which are parallel translates of each other. In other words, we regard two arrows which are parallel translates of each other as instances of the same vector. Given a vector v and a point p in the plane we can consider the particular arrow based at p which corresponds to v. We call its tip $p+v$. From this point of view each vector v defines an operation on the plane sending each point p to the point $p + v$. We call this operation translation through v, and denote it by $+v$ applied to the right so that the value of $+v$ acting on p is given in the usual way as $v + p$. Notice that $(p + v) + w = p + (v + w)$, or in other words the composition of the operation $+v$ with $+w$ is the operation $+(v+w)$.

Moreover, given any two points p and q there is a unique vector v for which $q = p + v$, namely the vector corresponding to the arrow from p to q. This is just another way of saying that a choice of origin p sets up a one-to-one correspondence between points and vectors, viz. q corresponds to the vector v which translates p to q.

The same structure appears in three-dimensional space. We have three-dimensional vectors v, represented by arrows, which define translation operations $+v$ satisfying $(p+v)+w = p+(v+w)$ for any two vectors v and w and any point p. Moreover, for any two points p and q there is a unique vector v such that $q = p + v$, so that choosing p as an origin sets up a one-to-one correspondence between points and vectors. A general affine space is defined as a set X and a vector space V, each vector v of which corresponds to a transformation $+v$ from X to itself called translation by v. The translations have to satisfy the two rules described above, that is, $(p+v)+w = p+(v+w)$ for any point p and vectors v and w, and given any two points of X there must be a unique translation that moves one to the other.

Affine spaces have a fundamental geometic significance in that they are to be considered flat, like the plane and three-dimensional space. A characteristic of affine spaces is the presence of special co-ordinate systems called affine co-ordinates. As we shall show, exponential families are affine spaces, and their canonical parameters are affine co-ordinates.

In the plane a pair of linearly independent arrows based at an origin determines a co-ordinate system. In vector space terms the arrows v_1, v_2 form a basis for the space of arrows, so that every arrow v can be expressed uniquely in the form $\theta_1 v_1 + \theta_2 v_2$. The numbers (θ_1, θ_2) can be regarded as co-ordinates for the point corresponding to v. If the arrows are of unit length and at right angles, and v_1, v_2 are in anticlockwise order, then such a co-ordinate system is a Cartesian co-ordinate system. In general these kinds of co-ordinate systems are called affine co-ordinates.

For a general affine space, having chosen an origin o we choose an ordered basis $v^1, v^2, \ldots, v^r$ for the space of translations. This is like choosing a set of axes at the origin. We obtain co-ordinates for points by expanding their corresponding vectors in terms of the basis. Each vector can be expressed as

$$v = \theta^1 v^1 + \theta^2 v^2 + \ldots + \theta^r v^r$$

The coefficients $\theta^i, i = 1, \ldots, r$ depend on v so we should really consider them as functions of v and write

$$v = \theta^1(v)v^1 + \theta^2(v)v^2 + \ldots + \theta^r(v)v^r$$

The functions $\theta^i, i = 1, \ldots, r$ are linear functions on V, and to regard them as co-ordinates of points of X amounts to composing them with the bijection $X \rightarrow V$ determined by the choice of origin o, i.e. identifying each point p in X with the corresponding v in V determined by o. We shall regard $\theta^i, i = 1, \ldots, r$ as functions on X in this way, as well as functions on V. Any collection of functions defined on X by such a process is called an *affine co-ordinate system.*

If θ and ϕ are any two affine co-ordinates systems on X then there is a matrix X^i_j and a vector $(u^1, \ldots, u^r)$, both depending on θ and ϕ, with

$$\theta^i(p) = \sum_{j=1}^{r} X^i_j \phi^j(p) + u^i$$

Indeed, $(u^1, \ldots, u^r)$ are the θ co-ordinates for the origin of the ϕ system, and X^i_j is the matrix which expresses the basis which produces ϕ in terms of the one which produces θ, i.e. the change of basis matrix. We say that two co-ordinate systems which are related in this way are *affinely related.*

Our interest lies in the fact that the converse is also true, namely if we have a set X and a collection of co-ordinates on X where any two are affinely related, then X is an affine space. From this we can infer that exponential families are affine spaces, since the canonical parameters form a collection of affinely related co-ordinate systems. However, the constructions involved in proving this fact are not very illuminating and we shall not detail them until Chapter 2. Instead we can reveal the affine structure of exponential families much more directly.

Example 1.3.1 Any vector space V is itself an affine space where the addition is just the usual addition of vectors.

Example 1.3.2 Let X be the set of all lines in $\mathbf{R}^2$ parallel to the y-axis. Define translation by v in $\mathbf{R}$ to mean translating the line x units. Then this makes X an affine space for $\mathbf{R}$.

Example 1.3.3 Let $X = (0, \infty)$ and $V = \mathbf{R}$. For x in X and v in $\mathbf{R}$ define

$$x + v = \exp(v)x$$

1.4 Log-likelihood and affine structure

The use of log-likelihoods amounts to treating sets of probability distributions as affine geometries. Many statisticians will already have an implicit sense of this relationship between geometry and statistics. Given two positive densities p_1 and p_2 we can regard multiplication by p_2/p_1 as translating the density p_1 into p_2. In general it is clear that multiplication of positive densities by positive functions gives a family of transformations of the set of positive densities parametrised by positive functions. However, the positive functions don't form a vector space. We can rectify this situation by considering an arbitrary function f to act on densities through multiplication by e^f. In other words, translation of the density p by the function f results in the density $e^f p$. Note that translation of p by f followed by translation by g results in the density $e^g e^f p = e^{(g+f)} p$, which is the same as its translation through $g + f$. Moreover, given any two densities p_1 and p_2 there is a unique function, namely $\log(p_2/p_1)$, which translates p_1 into p_2. Thus we should imagine densities to be like the points of a higher-dimensional plane, and functions to be like vectors or arrows which define translations shifting one density into another.

In order to give complete precision as well as complete generality to this discussion, we need to acknowledge that densities are densities with respect to some measure. When the underlying sample space is $\mathbf{R}^n$ the measure is usually Lebesgue measure dx. In any case, if Ω is a measure space we need to consider the families of non-negative measures which are absolutely continuous with respect to each other. To say that two such measures are absolutely continuous with respect to each other means that they have the same sets of measure zero. This is an equivalence relation, and under suitable regularity conditions it also implies that each has a density with respect to the other which is necessarily non-zero almost everywhere. The measures in any equivalence class form a family $\mathcal{M}$ which is naturally an affine space via the construction described above. Indeed we can let R_Ω be the vector space of measurable functions f on Ω and let translation by f be the operation sending the measure $d\mu$ into $e^f\,d\mu$.

Notice that expressing measures as densities with respect to any one of the measures a measure class $\mathcal{M}$ is exactly the process of choosing it as the origin of this affine space. If μ is the choice of origin we denote by

$$\ell: \mathcal{M} \to R_\Omega$$

the map $\ell(pd\mu) = \log(p)$. When p is a probability density $\ell(pd\mu)$ is called the *log-likelihood.* Although it is a slight abuse of terminology, we will call the map ℓ itself the log-likelihood. Thus treating measures as densities with respect to a base measure, and considering the log-likelihoods of these densities, amounts precisely to choosing an origin for $\mathcal{M}$ and identifying points of $\mathcal{M}$ with their translation vectors from the origin.

Denote by $\mathcal{P}$ the space of all probability measures in $\mathcal{M}$. Notice that probability measures cannot form an affine space inside $\mathcal{M}$, because multiplying them by functions e^f will almost certainly destroy the property that their total mass is one, and will very often destroy the property that their mass is even finite. However, rather than regard them as particular points in $\mathcal{M}$, probability measures can also be regarded as non-negative measures up to scale, because we can always divide a finite non-negative measure by its total mass to obtain a probability measure. Regarding a probability measure as a finite measure up to scale in effect treats it as an equivalence class of measures, with two measures being considered equivalent if they are rescalings of each other.

Notice that rescaling is one of the translation operations in our affine space of measures, namely translation by a constant function (multiplication by e^f with f constant). It follows that the measures in a measure class $\mathcal{M}$, when identified up to scale, form an affine space whose translation vectors are measurable functions f identified up to the addition of a constant. In effect, we treat the functions in the same way as we treat indefinite integrals; if two of them differ by a constant then one is as good as the other. In the case of indefinite integrals this is because two such functions have the same derivative. In the case of measures up to scale it is because two functions which differ by a constant will translate a measure into measures which differ only by scale. Hence measures up to scale form an affine space, and the set of probability measures

$\mathcal{P}$ is a subset of this affine space, namely the set corresponding to finite measures up to scale.

Example 1.4.1 *The binomial family.* Consider the simple case when $\Omega = \{1, 2\}$, a set with two elements. We can identify the space of random variables R_Ω with the $\mathbf{R}^2$ by associating to any random variable f its two values $(f(1), f(2)) = (f_1, f_2)$. Also we can identify $\mathcal{M}$ with the positive quadrant

$$\{(x, y) \mid x > 0, y > 0\}$$

by associating to a measure μ the pair $(\mu(\{1\}), \mu(\{2\})) = (\mu_1, \mu_2)$. The affine action of R_Ω on $\mathcal{M}$ is then

$$(\mu_1, \mu_2) + (f_1, f_2) = (\exp(f_1)\mu_1, \exp(f_2)\mu_2)$$

Changing a measure by scale, i.e. by translation through a constant function, produces the sets

$$\{(\exp(\lambda)\mu_1, \exp(\lambda)\mu_2) \mid \lambda \in \mathbf{R}\}$$

Therefore the equivalence classes of measures up to scale are precisely the lines through the origin intersected with the positive quadrant. This corresponds to the set of probability measures $\mathcal{P}$. The usual description of $\mathcal{P}$ is the line segment

$$\mathcal{P} = \{(\mu_1, \mu_2) \mid \mu_1 + \mu_2 = 1\}$$

which clearly intersects each radial line precisely once, confirming that a class of measures up to scale determines a unique probability distribution. In this more conventional description of $\mathcal{P}$, the affine action on $\mathcal{P}$ of functions identified up to the addition of a constant is

$$(\mu_1, \mu_2) + (f_1, f_2) = \frac{(\exp(f_1)\mu_1, \exp(f_2)\mu_2)}{\exp(f_1)\mu_1 + \exp(f_2)\mu_2}$$

A family P of measures up to scale which forms a finite-dimensional affine subspace is effectively an exponential family. An affine subspace is a subset which is closed under translation by a subspace of translation vectors. If this space is finite-dimensional, and is spanned by random variables $x^1, \ldots, x^r$, and if μ is one of the measures (up to scale), then the measures in such a family have the form

$$\mu + \theta^1 x^1 + \ldots + \theta^r x^r = \exp(\theta^1 x^1 + \ldots + \theta^r x^r - K)\mu$$

where changing K changes the scale of the measure. The probability measures in this subspace correspond to those which are finite, and we obtain the precise probability measure by choosing K to be $K(\theta)$, as determined by the formula

$$\log K(\theta) = \int_\Omega \exp(\theta^1 x^1 + \ldots + \theta^r x^r)\mu$$

Barndorff-Nielsen (1978) shows that the set of values of $\theta^1, \ldots, \theta^r$ for which these measures are finite form a convex set in $\mathbf{R}^r$, and such families of probability measures are called exponential families. Furthermore, the construction of such a family given above is manifestly the construction of an affine co-ordinate system for this finite-dimensional affine subspace, via a choice of origin and an ordered basis for the space of translations. This shows quite directly that the canonical co-ordinates $\theta^1, \ldots, \theta^r$ are an affine co-ordinate system for the affine geometry, and gives a deeper reason as to why all the canonical co-ordinate systems for an exponential family are affinely related. Specifically, our discussion shows that exponential families have a geometrical characterisation in terms of the finite-dimensional affine subspaces of measures up to scale, with their natural 'log-likelihood' affine structure.

Example 1.4.2 *The normal family as an exponential family.* Consider the normal family $\mathcal{N}$. Let V be the span of x^2 and x, U be the open set

$$U = \{ax^2 + bx \mid a > 0\},$$

and $\mu = dx$.

We write

$$\ell(x,p) = x^2\left(\frac{-1}{2\sigma^2}\right) + x\left(\frac{\mu}{\sigma^2}\right) - \log(\sqrt{2\pi}\sigma) - \frac{\mu^2}{2\sigma^2}$$

and deduce that the canonical co-ordinates are

$$\theta^1 = \frac{-1}{2\sigma^2} \quad \text{and} \quad \theta^2 = \frac{\mu}{\sigma^2}$$

and that

$$K(\theta) = \frac{1}{2}\log(\frac{-\pi}{\theta^1}) - \frac{(\theta^2)^2}{4\theta^1}$$

Example 1.4.3 *The unit normal family.* Let us call the family of all normal distributions with covariance matrix the identity matrix, the unit normal family, and denote it by $\mathcal{N}(\mathbf{R}^n, 1)$. Then the measures in $\mathcal{N}(\mathbf{R}^n, 1)$ are of the form

$$p(\theta)(x) = \frac{1}{(2\pi)^{n/2}}\exp(-\frac{|x-\theta|^2}{2})dx^1 \cdots dx^n$$

where we use the mean as the parameter θ. If we choose

$$\mu = \frac{1}{(2\pi)^{n/2}}\exp(-\frac{|x|^2}{2})dx^1 \cdots dx^n$$

as the origin the log-likelihood is

$$\ell(\theta)(x) = \exp(\sum x^i\theta^i - K(\theta))$$

where

$$K(\theta) = \frac{1}{2}|\theta|^2$$

So the θ co-ordinates are the canonical co-ordinates and if we identify $\mathcal{N}(\mathbf{R}^n, 1)$ with $\mathbf{R}^n$ using them the affine space structure is the usual affine space structure on $\mathbf{R}^n$.

In the next section we use the characterisation of an exponential family as the 'finite elements' of finite-dimensional affine subspaces of measures up to scale, to find a computable criterion for when a family of probability measures is exponential.

1.5 When is a family exponential?

1.5.1 *A geometric criterion*

Let

$$P = \{p(\theta)\}$$

be a parametrised family of probability measures which we will think of as a surface in $\mathcal{P} \subset \mathcal{M}$. We will not be precise about what it means for this surface to be differentiable but defer such questions to Chapter 3. Certainly we will assume that whenever we choose an origin and define the log-likelihood function $\ell(\theta)$ then this is a differentiable function of θ. Moreover we will assume that P has a well-defined tangent plane at every point p. This is an affine subspace of $\mathcal{M}$ which is tangent to P at p and hence has the form

$$\{p + v = \exp(v)p \mid v \in V\}$$

for some vector space of random variables V. We shall call the vector space V the tangent space to P at p and denote it by T_pP. We need to know one important fact about this tangent space; it has as a basis the score vectors

$$\frac{\partial \ell}{\partial \theta^i}(p)$$

for $i = 1, \ldots, r$. To see that the scores are in this space consider the curve

$$t \mapsto p(\theta^1, \ldots, \theta^i + t, \ldots, \theta^r)$$

defined by varying the ith parameter. At the point $p(\theta^1, \ldots, \theta^r)$ there is a vector v such that the affine line

$$t \mapsto \exp(tv)p(\theta^1, \ldots, \theta^r)$$

is tangent to this curve. If we choose an origin q we can define the log-likelihood by

$$p(\theta^1, \ldots, \theta^r) = \exp(\ell(\theta^1, \ldots, \theta^r))q$$

and applying this to both curves we deduce that

$$\ell(\theta^1, \ldots, \theta^i + t, \ldots, \theta^r)$$

must be tangent to

$$\ell(\theta^1, \ldots, \theta^r) + tv$$

By differentiating it follows that

$$v = \frac{\partial \ell}{\partial \theta^i}(p)$$

If we imagine that P is a surface in $\mathcal{M}$ then varying each of the parameters θ^i separately defines 'co-ordinate axes' on P. It is clear that whatever final definition we make in Chapter 3 of a statistical manifold we will want the tangents to these co-ordinate axes to span the tangent space at p. In other words we require that the scores span $T_p P$.

Let us return to consideration of the family P. Since we will want to deal with probability measures as positive measures up to scale it will be convenient to extend any such family to a family $\tilde{P}$ of positive measures defined by

$$\tilde{P} = \{\exp(\lambda)p \mid \lambda \in \mathbf{R}, p \in P\}$$

We extend the local co-ordinates θ^i by defining $\theta^i(\exp(\lambda)p) = \theta^i(p)$ and define a new co-ordinate by $\theta^0(\exp(\lambda)p) = \lambda$.

Let us denote by $\mathbf{R}.1$ the one-dimensional vector subspace of constant random variables. The 1 in this notation is the random variable which is everywhere equal to 1 and the $\mathbf{R}$ denotes that we want to consider all scalar multiples of 1, that is, the line in R_Ω containing 1. The space of random variables up to addition of constants is therefore the space $R_\Omega/\mathbf{R}.1$ of all lines in R_Ω in parallel to $\mathbf{R}.1$. The line through the random variable f is denoted $f + \mathbf{R}.1$ (see Figure 1.1) and addition and scalar multiplication of lines is defined by

$$a(f + \mathbf{R}.1) + b(g + \mathbf{R}.1) = (af + bg) + \mathbf{R}.1$$

The reader familiar with quotienting a vector space by a subspace will recognise, of course, that we are just discussing the quotient of the space of random variables by the subspace of constant random variables as the notation suggests.

It is easy to see that P is an affine subspace of $\mathcal{P}$ generated by a vector space $V \subset R_\Omega/\mathbf{R}.1$ if and only if $\tilde{P}$ is an affine subspace of $\mathcal{M}$ generated by

$$\tilde{V} = \{f \in R_\Omega \mid f + \mathbf{R}.1 \in V\}$$

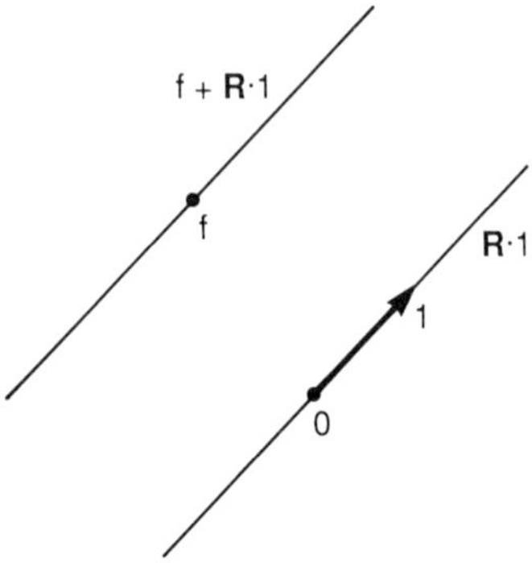

Figure 1.1

Hence P is an exponential family if and only if $\tilde{P}$ is an affine subspace of $\mathcal{M}$.

In order to conceive of a geometrical criterion for $\tilde{P}$ to be an affine subspace, let us observe the following property which all affine subspaces have. If μ is chosen as an origin for $\tilde{P}$ then every other measure in $\tilde{P}$ has the form $\exp(\tilde{\ell})\mu$ where $\tilde{\ell}$ ranges over the finite-dimensional subspace $\tilde{V}$ of random variables. Let $\tilde{\ell}(\exp(\lambda)p) = \ell(p) + \lambda$ be the log-likelihood generalised to this situation and let $\tilde{\ell}_i$ denote the ith partial derivative of $\tilde{\ell}$. Notice that by our definitions, $\tilde{\ell}_0$ is the constant random variable 1. Since all of the values of $\tilde{\ell}$ lie in the subspace $\tilde{V}$ so do the values of $\tilde{\ell}_0, \ldots \tilde{\ell}_r$ and moreover, as we have seen, they form a basis for $\tilde{V}$.

The key point which we shall use is that the partial derivatives of the score vectors must also lie in the space of translations $\tilde{V}$ when $\tilde{P}$ is affine. If we denote

$$\frac{\partial^2 \tilde{\ell}}{\partial \theta^i \partial \theta^j}$$

by $\tilde{\ell}_{ij}$ then, since $\tilde{\ell}_0, \ldots \tilde{\ell}_r$ is a basis for $\tilde{V}$, there must be coefficients $\gamma_{ij}^k(\theta)$ such that

$$\tilde{\ell}_{ij}(\theta) = \sum_{k=0}^{r} \gamma_{ij}^k(\theta) \tilde{\ell}_k(\theta)$$

Also as $\tilde{\ell}_{0i}(\theta) = 0$ and $\tilde{\ell}_0(\theta) = 1$ it follows that an equivalent way of stating this condition is that the second derivatives of the usual log-likelihood ℓ_{ij} are in the span of the usual scores ℓ_i and the constant random variables.

We claim that this property, that the derivatives of the scores lie in the span of the scores at each point, is characteristic of affine subspaces. It is not particularly difficult to prove this but we will defer the proof till Chapter 3. Instead let us consider an example.

Example 1.5.1 *The normal family.* As an example of the application of this criterion let us consider how to show that the normal family is an exponential family. We have

$$\ell(\mu,\sigma) = -\frac{(x-\mu)^2}{2\sigma^2} - \log(\sqrt{2\pi}\sigma)$$

and hence the scores are

$$\ell_1 = \frac{(x-\mu)}{\sigma^2} \quad \text{and} \quad \ell_2 = \frac{(x-\mu)^2}{\sigma^3} - \frac{1}{\sigma}$$

The non-zero second partial derivatives are

$$\ell_{11} = \frac{-1}{\sigma^2}$$

$$\ell_{12} = \ell_{21} = \frac{-2(x-\mu)}{\sigma^3}$$

and

$$\ell_{22} = \frac{-3(x-\mu)^2}{\sigma^4} + \frac{1}{\sigma^2}$$

From these formulae we can see that

$$\begin{aligned}
\ell_{11}(\mu,\sigma) &= \frac{-1}{\sigma^2}.1 \\
\ell_{12}(\mu,\sigma) &= \frac{-2}{\sigma}\ell_1(\mu,\sigma) \\
\ell_{22}(\mu,\sigma) &= \frac{-3}{\sigma}\ell_2(\mu,\sigma) - \frac{2}{\sigma^2}.1
\end{aligned}$$

while the conditions for $\ell_{0j}(\mu,\sigma)$ are satisfied automatically since these are zero.

1.5.2 *A computable criterion*

For the normal family it is easy to find an expansion of the $\tilde{\ell}_{ij}$ in terms of the scores by inspection. In other examples it may not be straightforward to find such an expansion and in some examples we may want to prove that no such coefficients exist. Let us use this criterion to devise a method of determining precisely when a family is exponential. As we have seen it suffices to know if the ℓ_{ij} are in the tangent space to $\tilde{P}$ for each $i, j = 1, \ldots, r$. Let E_p denote the expectation of functions with respect to the measure p. We define an inner product by the formula

$$\langle f, g \rangle_p = E_p(fg)$$

on the subspace of p square-integrable random variables f, i.e. those random variables satisfying $E_p(f^2) < \infty$. For convenience let us assume that all random variables satisfy this condition. In general it suffices to assume that the scores at p and the second derivatives $\tilde{\ell}_{ij}(p)$ are p square-integrable. The matrix

$$g_{ij}(p) = E_p(\ell_i \ell_j)$$

is called the *Fisher information matrix.* If we restrict the inner product $\langle \ , \ \rangle_p$ to $T_p P$ it defines an inner product on this tangent space and the Fisher information matrix is just the matrix of this inner product with respect to the basis defined by the scores.

This inner product on R_Ω defines a normal space N_p to $T_p\tilde{P}$ such that

$$R_\Omega = N_p \oplus T_p\tilde{P}.$$

It is straightforward to check that if f is a random variable its normal component in N_p is

$$\pi_p(f) = f - \sum_{m,n} g^{mn} E_p(f\ell_m)\ell_n - E_p(f)$$

where g^{mn} is the inverse of the Fisher information matrix. Indeed

$$\langle \pi_p(f), 1 \rangle = E_p(f) - \sum g^{mn} E_p(f\ell_m) E_p(\ell_n) - E_p(f) = 0$$

where we use the fact that $E_p(\ell_i) = 0$ for any i, which can be proved by differentiating

$$1 = \int_\Omega \exp(\ell) p$$

Also

$$\begin{aligned}\langle \pi_p(f), \ell_i \rangle =& E_p(f\ell_i) - \sum g^{mn} E_p(f\ell_m) E_p(\ell_n \ell_i) \\ =& E_p(f\ell_i) - \sum g^{mn} g_{ni} E_p(f\ell_m) \\ =& 0\end{aligned}$$

This proves that $\pi_p(f)$ is normal to $T_p\tilde{P}$. Moreover we can write

$$f = (f - \pi_p(f)) + \pi_p(f)$$

and $f - \pi_p(f)$ is a linear combinations of the scores and the constant random variables so in $T_p\tilde{P}$. It follows that $\pi_p(f)$ is the component of f normal to $T_p\tilde{P}$.

We have proved that the family is exponential if the functions ℓ_{ij} are always tangential to $\tilde{P}$. This is equivalent to the normal component of each ℓ_{ij} vanishing, that is, to

$$\alpha_{ij} = \pi_p(\ell_{ij}) = \ell_{ij} - \sum_{mn} g^{mn} E(\ell_{ij}\ell_m))\ell_n - E(\ell_{ij}) = 0$$

This quantity α_{ij} we shall call the *second fundamental form* of the family. It is also called the imbedding curvature in Amari (1985).

The name second fundamental form arises classically in the theory of parametrised surfaces. Consider for instance a parametrised surface

$$\Sigma = \{x(\theta^1, \theta^2)\} \subset \mathbf{R}^3$$

The first fundamental form is defined to be

$$g_{ij} = \langle \frac{\partial x}{\partial \theta^i}, \frac{\partial x}{\partial \theta^j} \rangle$$

where $\langle\ ,\ \rangle$ is the usual inner product on $\mathbf{R}^3$ and is analogous to the Fisher information. It has the same geometric interpretation as defining an inner product on the tangent plane to Σ. The second fundamental form is the component of the second derivative of x perpendicular to the tangent plane to Σ. Because the tangent plane to Σ is spanned by the vectors

$$\frac{\partial x}{\partial \theta^1} \quad \text{and} \quad \frac{\partial x}{\partial \theta^2}$$

the second fundamental form of Σ is

$$\alpha_{ij} = \frac{\partial^2 x}{\partial\theta^i \partial\theta^j} - \sum_{mn} g^{mn} \langle \frac{\partial^2 x}{\partial\theta^i \partial\theta^j}, \frac{\partial x}{\partial\theta^m} \rangle \frac{\partial x}{\partial\theta^n}$$

The second fundamental form has the same interpretation in this case, namely the surface Σ is a plane if and only if the second fundamental form vanishes.

The geometric idea behind the fact that the second fundamental form vanishing characterises affine subspaces is the following. We think of the scores $\tilde{\ell}_i$ as tangent vectors to $\tilde{P}$. As we move around on $\tilde{\mathcal{P}}$ these vectors change; this is what $\tilde{\ell}_{ij}$ is measuring. These changes are to be regarded as due to two causes. The first cause is that the lines in $\tilde{\mathcal{P}}$ on which the parameters are constant may be bending around and the second cause is that $\tilde{\mathcal{P}}$ may be bending around inside $\mathcal{M}$. The tangential and normal components of $\tilde{\ell}_{ij}$ measure these two types of bending respectively. So the vanishing of the normal component of $\tilde{\ell}_{ij}$ corresponds to the fact that the surface $\tilde{P}$ is not bending.

We can refine this condition for a family to be exponential by replacing α_{ij} by a scalar quantity γ which vanishes if and only if α_{ij} does. The definition of γ is

$$\gamma(p) = \sum_{ijkl} g^{ij}(p) g^{kl}(p) E_p(\alpha_{ik}(p)\alpha_{jl}(p))$$

and because $g^{ij}(p)$, the inverse of the Fisher information matrix, is postive definite we see that γ vanishes precisely when α_{ij} vanishes. The function γ, in the case of a one-dimensional family is Efron's statistical curvature Efron (1975).

1.6 Parameter independence

In this section we want to discuss precisely how α_{ij} depends on the choice of co-ordinates. Note that we have shown that four things are equivalent to a family being exponential. The first is the condition that a subset of the space of probability measures be affine. This is manifestly independent of the parameters. The second is the condition that the second derivatives of the log-likelihood are in the span of the scores and the constants. It is not clear that this is independent of parameters but by the equivalence we have established it must be. Thirdly there is the vanishing of all the α_{ij}. This is also not manifestly independent of co-ordinates but again the equivalence shows that it must be. Finally there is the vanishing of γ. We shall see that γ is a function on P which is independent of the choice of parameters.

Let us confirm this co-ordinate independence by calculation. Consider first the question of the second derivatives being in the span of the scores and the constant functions. We have for a choice of parameters θ that

$$\frac{\partial^2 \ell}{\partial \theta^i \partial \theta^j} = \gamma_{ij}^k \frac{\partial \ell}{\partial \theta^k} + \gamma_{ij} \tag{1.6.1}$$

Here and throughout this book we will often adopt the useful 'Einstein summation convention' that any index which occurs both raised and lowered is summed over. If we change to parameters χ^i then we have, by the chain rule,

$$\frac{\partial \ell}{\partial \chi^i} = \frac{\partial \theta^k}{\partial \chi^i} \frac{\partial \ell}{\partial \theta^k} \tag{1.6.2}$$

and

$$\frac{\partial^2 \ell}{\partial \chi^i \partial \chi^j} = \frac{\partial^2 \ell}{\partial \theta^k \partial \theta^l} \frac{\partial \theta^k}{\partial \chi^i} \frac{\partial \theta^l}{\partial \chi^j} + \frac{\partial^2 \theta^l}{\partial \chi^i \partial \chi^j} \frac{\partial \ell}{\partial \theta^l}. \tag{1.6.3}$$

Substituting (1.6.1) and (1.6.2) into (1.6.3) gives

$$\frac{\partial^2 \ell}{\partial \chi^i \partial \chi^j} = (\gamma_{nl}^m \frac{\partial \chi^k}{\partial \theta^m} \frac{\partial \theta^n}{\partial \chi^i} \frac{\partial \theta^l}{\partial \chi^j} + \frac{\partial^2 \theta^k}{\partial \chi^i \partial \chi^j} \frac{\partial \chi^k}{\partial \theta^l}) \frac{\partial \ell}{\partial \chi^k} + \gamma_{lm} \frac{\partial \theta^m}{\partial \chi^i} \frac{\partial \theta^l}{\partial \chi^j}$$

so if

$$\frac{\partial^2 \ell}{\partial \theta^i \partial \theta^j}$$

is in the the span of the scores and the constant random variables so also is

$$\frac{\partial^2 \ell}{\partial \chi^i \partial \chi^j}$$

To see that the vanishing of the α_{ij} is independent of the choice of co-ordinates assume as above that χ is another parametrisation. We shall distinguish between it and θ by letting θ co-ordinates have indices $i, j, \ldots$ and χ co-ordinates have indices $a, b, \ldots$. Then as we have seen

$$\frac{\partial^2 \ell}{\partial \chi^a \partial \chi^b} = \frac{\partial^2 \ell}{\partial \theta^i \partial \theta^j} \frac{\partial \theta^i}{\partial \chi^a} \frac{\partial \theta^j}{\partial \chi^b} + \frac{\partial^2 \theta^k}{\partial \chi^a \partial \chi^b} \frac{\partial \ell}{\partial \theta^k}$$

It is possible to substitute this into the expression for α_{ij} and calculate away. However it is simpler to recall that α_{ij} is the projection of the second derivative of ℓ onto the orthogonal space to $T_p\tilde{P}$. Then

$$\begin{aligned} \alpha_{ab} &= \pi\left(\frac{\partial^2 \ell}{\partial \chi^a \partial \chi^b}\right) \\ &= \pi\left(\frac{\partial^2 \ell}{\partial \theta^i \partial \theta^j} \frac{\partial \theta^i}{\partial \chi^a} \frac{\partial \theta^j}{\partial \chi^b}\right) \\ &= \alpha_{ij} \frac{\partial \theta^i}{\partial \chi^a} \frac{\partial \theta^j}{\partial \chi^b} \end{aligned}$$

using the fact that

$$\pi\left(\frac{\partial \ell}{\partial \theta^i}\right) = 0$$

It follows that α_{ab} vanishes precisely when α_{ij} vanishes.

Finally if we consider γ we have to first note that if

$$g_{ab} = E_p\left(\frac{\partial^2 \ell}{\partial \chi^a \partial \chi^b}\right)$$

is the Fisher information with respect to the co-ordinates χ then we have

$$g_{ij} = \frac{\partial \chi^a}{\partial \theta^i} \frac{\partial \chi^b}{\partial \theta^j} g_{ab}$$

It follows readily that γ is a function on P that is independent of the choice of co-ordinates. That is, it is an *invariant* of the family.

There are two approaches that can be taken to invariants such as γ. The straightforward approach is to just define them in co-ordinates and check that if the co-ordinates are changed then they do not change. Hey presto! we have an invariant. Unfortunately this approach sheds no light on why the quantity in question is invariant or what is the statistical significance of its invariance. Moreover if the calculation takes ten pages or needs a computer this approach starts to become unwieldy. Our approach is to give a geometric meaning to a quantity such as γ which is manifestly independent of any choice of co-ordinates. The effort of calculation is replaced by the effort of learning some geometric concepts. However the dividend is that we can then hope to extract statistical significance from geometric significance. For instance, without geometry as a guide it would be hard to imagine why one would propose the vanishing of γ as a criterion for the exponentiality of a family.

Along the way we will also discover the geometric significance of quantities such as the score and the Fisher information which are not invariant but nevertheless seem to transform in an orderly fashion when we change co-ordinates.

1.7 Remarks for Chapter 1

Remark 1.1 Sometimes an exponential family is presented in the apparently more general form

$$p(x,\theta) = \exp(\sum_{i=1}^{r} x^i\theta^i - K(\theta) - \eta(x))$$

However if we replace the measure $d\mu$ by the measure $\exp(-\eta)d\mu$ then the densities are all in the form we have considered and we see that no generality has really been sacrificed.

Remark 1.2 Much of what we have said here about exponential families is well known to statisticians, although perhaps not in the precise form we give it here. For those familiar with the standard reference Barndorff-Nielsen (1978) let us make two remarks.

The first is that in our setting it is not assumed that the sample space is a subset of $\mathbf{R}^n$. Of course, as we aren't trying to do anything technical with our measures we don't need to make this assumption. Moreoever once canonical co-ordinates are chosen as in equation (1.2.1) the random variables, considered as a map $x: \Omega \to \mathbf{R}^r$, are a sufficient statistic (see Chapter 7) and we could therefore replace Ω by its image under x.

The second comment is that in Barndorff-Nielsen (1978) notions of convexity for subsets of Euclidean space, play an important role. All these notions can also be applied to a finite-dimensional affine space. Given any two points p and q in an affine space there is a vector v such that $q = p + v$. The line segment joining p and q is then defined to be the set

$$[p, q] = \{p + tv \mid t \in [0, 1]\}$$

and a set is said to be convex if whenever it contains two points it contains the line segment joining them.

Remark 1.3 Readers interested in the second fundamental form for surfaces and submanifolds of Riemannian manifolds should consult Klingenberg (1978) or Koboyashi and Nomizu (1963).

1.8 Exercises for Chapter 1

Exercise 1.1 Derive the relations between the (μ, σ), canonical and expectation parameters implied by requiring that

$$p(\mu, \sigma) = p(\theta^1, \theta^2) = p(\eta^1, \eta^2)$$

Show that

$$\mathcal{N} = \{p(\theta^1, \theta^2)dx \mid \theta^2 \in \mathbf{R}, \theta^1 < 0\}$$

and

$$\mathcal{N} = \{p(\eta^1, \eta^2)dx \mid \eta^2 \in \mathbf{R}, \eta^1 > 0\}$$

Here $\mathcal{N}$ is the set of normal densities defined in section 1.1.

Exercise 1.2 If p is a probability measure and f is a random variable let

$$p + f = \frac{\exp(f)}{E_p(\exp(f))}p$$

Prove that $(p + f) + g = p + (f + g)$.

Exercise 1.3 Let $\mathcal{N}(\mathbf{R}^n)$ denote that family of all normal distributions on $\mathbf{R}^n$. The measures in $\mathcal{N}(\mathbf{R}^n)$ are parametrised by the mean $\mu = (\mu_1, \ldots, \mu_n)$ and the covariance matrix $\Sigma = (\Sigma_{ij})$, a symmetric, positive definite matrix. Their explicit form is

$$p(\mu, \Sigma)(x) = \frac{|\det(\Sigma)|^{n/2}}{(2\pi)^{n/2}} \exp(-\frac{1}{2}(x - \mu)\Sigma(x - \mu)^t)dx^1 \cdots dx^n$$

Show that $\mathcal{N}(\mathbf{R}^n)$ is an exponential family whose dimension is $(1/2)n(n + 1)$.

CHAPTER 2

Calculus on manifolds

2.1 Introduction

In Chapter 1 we introduced the idea of freeing a parametric family of probability distributions from its parameters, and treating the family as a geometric object like a surface. In this chapter we shall pursue the same idea in respect of differential calculus, and show how the theory of rates of change of functions can be freed from the notion of dependent and independent variables. However, co-ordinate systems or parametrisations, which are choices of independent variables, are not totally dispensable. They impose a certain minimal level of geometric structure on the set which they parametrise, in that they collectively determine the notion of tangent vector which is the foundation for describing rates of change of functions.

This geometric approach to calculus is called calculus on manifolds. In succeeding chapters we will show how differential geometry amounts to an extension of this program to second derivatives, while the asymptotic expansions common in statistics require extensions to even higher order. It is as a geometric theory of higher-order rates of change of functions that differential geometry enters fundamentally into statistical theory.

2.2 The basic apparatus

2.2.1 *Functions, variables and parameters*

Calculus on manifolds is a theory of rates of change which deals directly with functions defined on abstract sets rather than with functions of several variables. As an example of this distinction, consider a temperature distribution on a plane. The plane is an abstract set of points, and the temperature distribution is a function on this set whose value at any point is the temperature at that point. The temperature function employs no co-ordinates or independent variables for its description, and its evaluation at any point occurs in principle by some process of measurement rather than by any analytical manipulations with co-ordinates. It is not *a priori* a function of several variables, but only becomes one by choosing some co-ordinate system on the plane, such as Cartesian or polar co-ordinates, to play the role of independent variables.

As a more definite example, consider the function on the plane given by the square of the distance of any point from a chosen origin. Given any point in the plane one evaluates the function at that point again by a process of measurement rather than the use of any co-ordinate formulas. However, it can be expressed as a function of two variables by introducing, say, a Cartesian co-ordinate system (x, y) with this origin, whereupon it is given by $x^2 + y^2$. Alternatively it can be expressed by r^2 as a function of polar co-ordinates (r, θ) with the same origin. This illustrates the fact that, in general, a given function on the plane must be expressed as different functions in the variables of different co-ordinate systems. Our philosophy will be that one should work with the actual functions on the plane, and not with the various forms of functions which they assume under various co-ordinate systems or parametrisations of the plane by so-called independent variables. Calculus on manifolds is essentially an approach to calculus which is consistent with this philosophy.

Rather than dispense with parameters, independent variables, co-ordinates and such like, calculus on manifolds recognises that these are simply examples of the very functions to which the theory must apply, and often very important examples at that. For instance, the x and y co-ordinates of a Cartesian co-ordinate system are actually functions on the plane. Indeed, for each point p in the plane there are numbers called its x and y co-ordinates, which we should really denote by $x(p)$ and $y(p)$ in order to indicate their dependence upon the point p. While the notations $x(p)$ and $y(p)$ are rarely used despite their correctness because they are somewhat cumbersome, they have the advantage of reminding us that x and y are functions which are to be evaluated at any point of the plane by measuring its perpendicular distances from the axes of the Cartesian system. As functions on the plane, concepts such as the rate of change of a function should apply to x and y in the same way as they apply to other functions such as the temperature or square-of-distance discussed above. This point of view stands in contrast to several-variable calculus, where x and y are given a special role as independent variables in terms of which other functions are expressed, and their rates of change interpreted through the changes induced by variations in the values of x and y. Within such a framework the rates of change of x and y themselves appear somewhat nonsensical or at best tautological.

In particular we shall treat the parameters $\theta^1, \ldots, \theta^n$ of a family P of probability distributions as functions on the set P. Instead of focusing on the role of parameters as specifying particular probability distribution, we emphasise the alternative point of view in which each probability distribution has a list of parameters corresponding to it. From this point of view the parameters depend upon, are determined by, or are functions of, the probability distribution in the set P. For example, we treat μ and σ as quantities determined by the distributions of the normal family, and hence as functions on the set of all normal distributions.

The practice of regarding variables and parameters as functions on abstract sets has not arisen for the sake of statistics: it is quite commonplace in physics. For example, the pressure and temperature of a given mass of gas are considered to determine a unique thermodynamic state of the gas, whereby pressure and temperature parametrise the set of all such thermodynamic states. One therefore conceives of an abstract set of thermodynamic states of the gas, with pressure and temperature as functions on the set forming a co-ordinate system on it. The pressure and volume of the gas also form such a co-ordinate system. In the theory of relativity one considers an observer's measurements of time and of space co-ordinates as together determining a unique 'event' , so that one conceives of the cosmos as a set of events with each observer's time and space co-ordinates together forming a four-dimensional co-ordinate system on it. The point of relativity theory is that each observer creates an *a priori* different co-ordinate system for the set of all events in the cosmos. The principle that the laws of physics should not give preference to any particular parametrisations has given great impetus to the formulation of the concepts of calculus in terms of the underlying abstract sets of states or events. In statistics as well, we shall pursue the goal of formulating statistical theory in terms of the underlying set of probability distributions, rather than in terms of convenient parametrisations of it.

We now give some examples of naturally defined functions that are functions of probability distributions, not *a priori* functions of many variables.

Example 2.2.1 *The log-likelihood.* If we think of a collection of probability distributions P as an abstract set then we can define a family of real-valued functions on it as follows. We choose an origin μ for the space of measures and define the log-likelihood function

$$\ell: P \to R_\Omega$$

by requiring that

$$p = \exp(\ell(p))\mu$$

If we choose a sample point x in Ω then

$$\ell_x(p) = \ell(p)(x)$$

is a real-valued function on P. The important point is that this function depends only on the choice of μ and x and, moreover, the argument of this function is not a collection of numbers but a probability distribution p.

Example 2.2.2 *The normal family.* As usual we let

$$\mathcal{N} = \{\frac{1}{\sqrt{2\pi}\sigma}\exp(\frac{-(x-\mu)^2}{2\sigma^2})dx\}$$

We define two functions μ and σ on this set by

$$\mu(p) = \int_{\mathbf{R}} xp$$

and

$$\sigma(p) = \Big(\int_{\mathbf{R}} (x^2)p - \big(\int_{\mathbf{R}} xp\big)^2\Big)^{1/2}$$

These functions are, of course, the mean and standard deviation of the measure p. Again we want to stress the fact that these are functions of probability distributions. Of course they are a particularly special pair of functions: they define co-ordinates on $\mathcal{N}$. This is a consequence of the fact that the function

$$\begin{array}{ccc} \mathcal{N} & \to & \{(\mu,\sigma) \mid \sigma > 0\} \\ p & \mapsto & (\mu(p), \sigma(p)) \end{array}$$

is one to one and onto, that is, it is a bijection.

2.2.2 *Rates of change under variations*

The most rudimentary notion of the rate of change of a given function f on an abstract set is its rate of change at a point p under a *variation* of the point. By a variation of a point p in a set P we mean a map $\gamma\colon(-\epsilon,\epsilon) \to P$, where $(-\epsilon,\epsilon)$ is an interval of numbers, and where $\gamma(0) = p$. One can think of a variation as a moving point in P, with $(-\epsilon,\epsilon)$ as a 'time' interval and $\gamma(t)$ giving the location of the point at each 'time' $t \in (-\epsilon,\epsilon)$. In particular, the 'timing' is arranged so that at time 0 the moving point passes through the point of interest p. Given any function f on P we can consider its value at the moving point as a function of time. In other words we consider the function whose value at $t \in (-\epsilon,\epsilon)$ is $f(\gamma(t))$. See Figure 2.1.

This function is properly denoted $f \circ \gamma$, although classically it is often written $f(t)$, omitting the γ. For example, a moving point in the plane is often described $(x(t), y(t))$, whereas it is more properly described $(x \circ \gamma, y \circ \gamma)$, since $x(t)$ and $y(t)$ are intended to be the Cartesian co-ordinates of some point $\gamma(t)$ in the plane whose x and y co-ordinates should therefore be written $x(\gamma(t))$ and $y(\gamma(t))$.

The image of a variation γ can generally be thought of as a curve, although there are exceptional cases. For example, the image of the constant variation given by $\gamma(t) = p$ for every t is the single point p. The image of a variation can be interpreted as the trace left by the moving point which the variation describes. However, it is the motion of the point that will concern us and not merely the trace

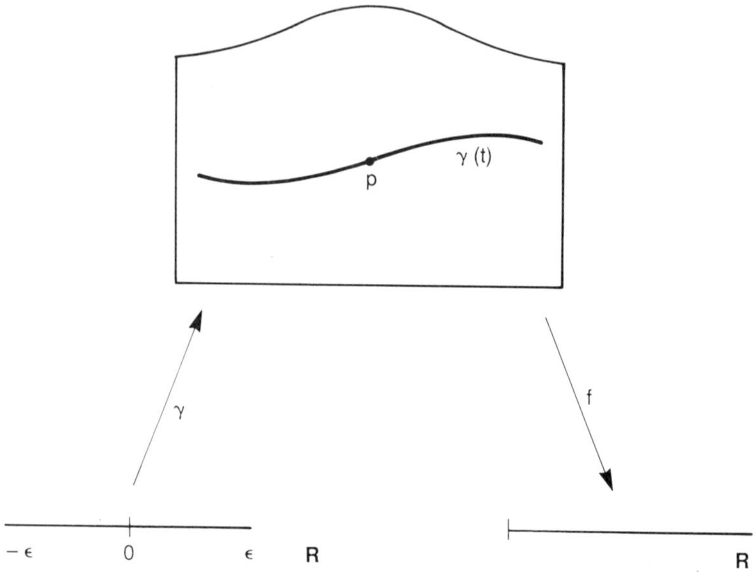

Figure 2.1

which it leaves, so that variations are in no way intended merely to represent curves via parametrisations which can be ignored. It will matter to us, for example, if some point is moving along the same trace twice as fast as another, and this will show up as a difference in the maps γ, even though they leave the same trace.

The rate of change of a function f under a variation γ through p is defined by

$$\left.\frac{d}{dt}f(\gamma(t))\right|_{t=0}$$

A more compact notation is $(f\circ\gamma)'(0)$. Note that we are concerned with the rate of change of the function f only at the point p, which is why we consider the derivative of $f\circ\gamma$ only at 0.

The concept of the rate of change of a function under a variation includes as a special case the concept of its partial derivatives with respect to a co-ordinate system. Any co-ordinate system on a set determines a special collection of variations or motions through each point of a set. These are the variations for which all the co-ordinate functions are held constant except one, which changes at a unit rate. Specifically, if $\phi^1, \ldots, \phi^n$ is a co-ordinate system on a set P and $p \in P$ then we let $\gamma_i(h)$ be the element of P such that

$$\begin{aligned} \phi^1(\gamma_i(h)) &= \phi^1(p), \\ &\vdots \\ \phi^i(\gamma_i(h)) &= \phi^i(p) + h, \\ &\vdots \\ \phi^n(\gamma_i(h)) &= \phi^n(p). \end{aligned}$$

We call γ_i the ith co-ordinate variation through p. (See Figure. 2.2). These are exactly the variations used to define the partial derivatives of f with respect to ϕ^i, and indeed we define

$$\frac{\partial f}{\partial \phi^i} := (f \circ \gamma_i)'(0)$$

For example, consider the definition of $\partial f/\partial x$, where f is a function on a plane, and (x, y) is a Cartesian co-ordinate system. At a point p with co-ordinates (x_0, y_0) it is usually described as

$$\frac{\partial f}{\partial x} := \lim_{h \to 0} \frac{f(x_0 + h, y_0) - f(x_0, y_0)}{h}$$

However, this formula needs interpretation since, for a start, the argument of f should be a point of the plane and not some pair of numbers such as (x_0, y_0). We can explain everything properly by introducing the function $\bar{f}$ defined on pairs of numbers by the relation

$$\bar{f}(x(q), y(q)) := f(q)$$

for each point q in the plane (see Figure 2.3).

(Of course, $\bar{f}$ has nothing to do with any kind of sample mean, and we apologise that our notation might trigger any deeply ingrained temptation to interpret it so.)

The precise definition of $\partial f/\partial x$ at p is

$$\frac{\partial f}{\partial x} := \lim_{h \to 0} \frac{\bar{f}(x_0 + h, y_0) - \bar{f}(x_0, y_0)}{h}$$

Now let γ_x be the x co-ordinate variation, so that $\gamma_x(h)$ is the point with co-ordinates $(x_0 + h, y_0)$. Notice that γ_x is a variation through p, because $\gamma_x(0)$ is the point with co-ordinates (x_0, y_0) which is p,

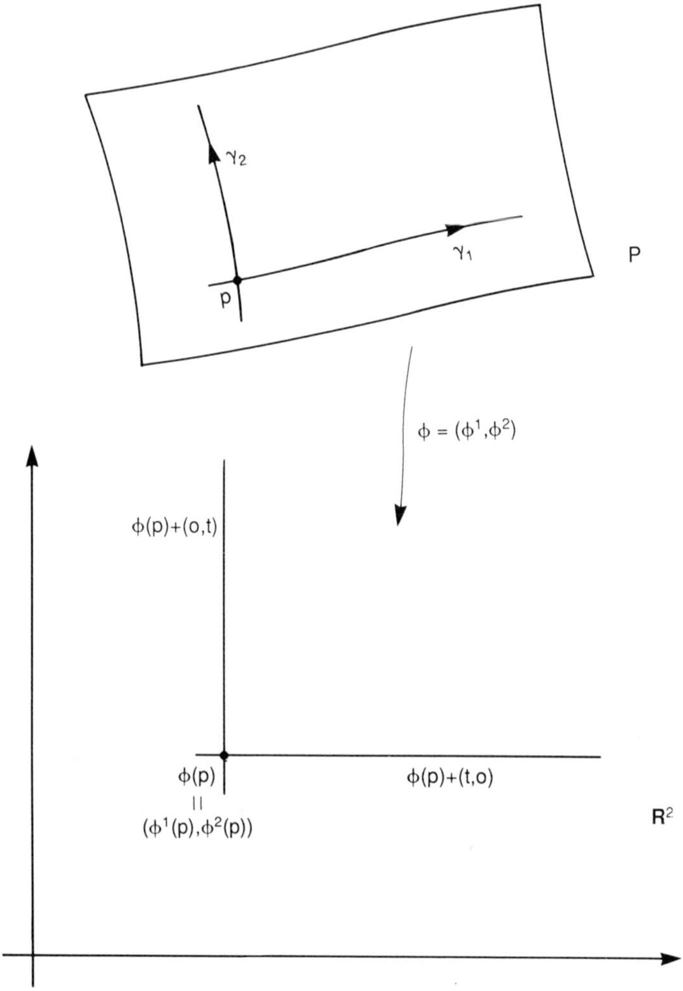

Figure 2.2

and it describes a motion in which y is held constant but x changes at unit rate. By definition we have

$$\bar{f}(x_0 + h, y_0) - \bar{f}(x_0, y_0) = f(\gamma_x(h)) - f(\gamma_x(0))$$

It follows that the limit which defines $\partial f/\partial x$ at p is just $(f \circ \gamma_x)'(0)$,

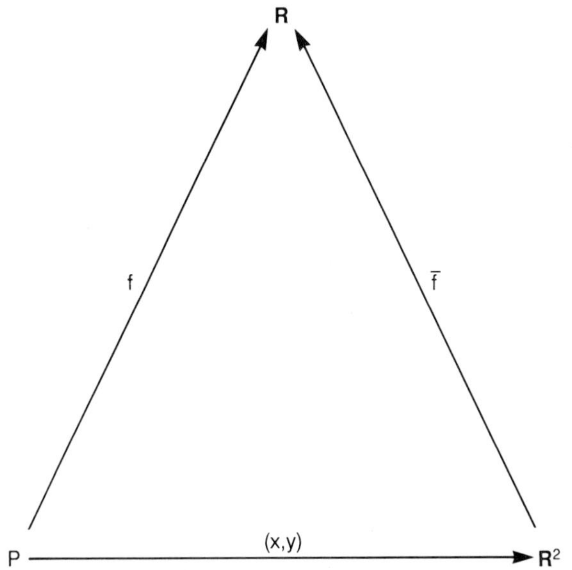

Figure 2.3

which is the derivative of f under the co-ordinate variation γ_x. See Figure 2.4.

If $\phi^1, \ldots, \phi^n$ is a co-ordinate system on a more general set P, and f is any function on P, then we can define a function $\bar{f}$ by the formula

$$f(p) = \bar{f}\left(\phi^1(p), \ldots, \phi^n(p)\right) \tag{2.2.1}$$

for every point p in the set P. Note that whereas f is a function on the set P, the domain of $\bar{f}$ is some subset of $\mathbf{R}^n$. As is often the case with identities between functions, the argument p can be left out and (2.2.1) written

$$f = \bar{f}\left(\phi^1, \ldots, \phi^n\right)$$

(2.2.1) is called the co-ordinate expression of f, and functions on a set are commonly defined by giving their co-ordinate expressions in some co-ordinate system. For example, one is often invited to consider the function $x^2 + y^2$ on the plane. Properly interpreted, it is the function f on the plane whose value at a point p is determined in two stages, first by evaluating the functions x and y at p, that is,

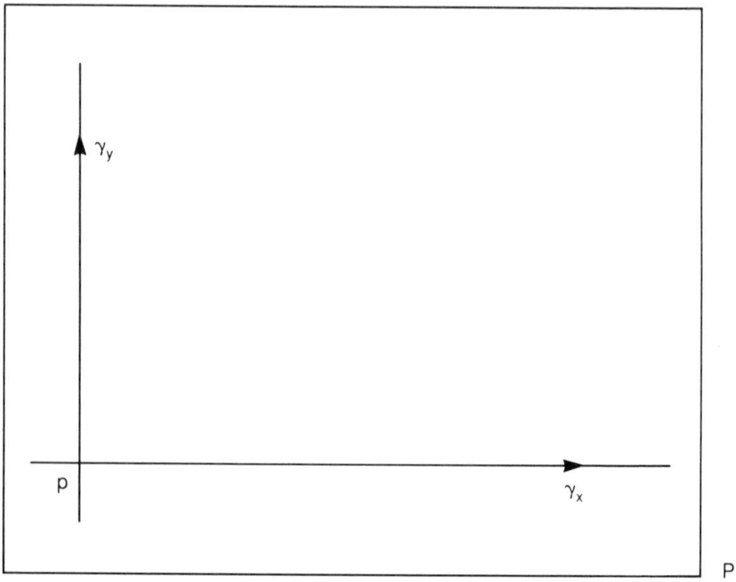

Figure 2.4

by determining the co-ordinates of p, and then taking the sums of the squares of the resulting pair of numbers $(x(p), y(p))$. This last step, which is the one of greatest interest, is a purely arithmetic operation which applies to any pair of numbers. This is actually the function $\bar{f}$. If one regards x and y in the formula $x^2 + y^2$ as a pair of numbers, rather than as functions on the plane, then the formula describes $\bar{f}$ rather than f. It is an extremely convenient ambiguity, present in all co-ordinate expressions for functions on a set, that formulas in co-ordinates describe the function itself if the co-ordinates are interpreted as functions on the set, while they describe the function $\bar{f}$ if the co-ordinates are interpreted as numbers. It is of fundamental importance that the reader be aware of this ambiguity. Notice that this ambiguity also carries over to the description of the partial derivatives of a function.

The definition of $\bar{f}$ begs the question as to the precise conditions under which it exists. These are simply that the co-ordinates ϕ define a one-to-one correspondence between the subset U of points of P that they co-ordinatise, and the set $\phi(U)$ of actual co-ordinate values in $\mathbf{R}^n$. In this case $\bar{f}$ is defined on $\phi(U)$.

Example 2.2.3 *The normal family.* Consider again the example of the normal family

$$\mathcal{N} = \{\frac{1}{\sqrt{2\pi}\sigma}\exp(\frac{-(x-\mu)^2}{2\sigma^2})dx\}$$

It is straighforward to check that the log-likelihood function and the mean and standard deviation functions are related by

$$\ell_x = \frac{-(x-\mu)^2}{2\sigma^2} - \log(\sqrt{2\pi}\sigma)$$

We have seen that the mean and standard deviation also provide us with co-ordinates on $\mathcal{N}$ so there is a function

$$\bar{\ell}_x : \{(\mu, \sigma) \mid \mu > 0\} \to \mathbf{R}$$

of two variables defined by $\ell(p) = \bar{\ell}(\mu(p), \sigma(p))$. This function is, of course,

$$\bar{\ell}_x(\mu, \sigma) = \frac{-(x-\mu)^2}{2\sigma^2} + \log(\sqrt{2\pi}\sigma)$$

We can now see the ambiguity that we have just discussed. The formula

$$\bar{\ell}_x(\mu, \sigma) = \frac{-(x-\mu)^2}{2\sigma^2} + \log(\sqrt{2\pi}\sigma)$$

defines a function on $\{(\mu, \sigma) \mid \mu > 0\}$ and the formula

$$\ell_x = \frac{-(x-\mu)^2}{2\sigma^2} - \log(\sqrt{2\pi}\sigma)$$

is an equation relating three functions on $\mathcal{N}$.

2.2.3 *The chain rule, velocities and tangent vectors*

The derivative of a function under a variation involves only one-dimensional calculus and *a fortiori* the same is true of partial derivatives. The first genuinely multi-dimensional result is the chain rule. If γ is any path through a point p in a set P, and $\phi^1, \ldots, \phi^n$ is a co-ordinate system, then we can formulate the chain rule as

$$(f \circ \gamma)'(0) = \frac{\partial f}{\partial \phi^1}(p)(\phi^1 \circ \gamma)'(0) + \ldots + \frac{\partial f}{\partial \phi^n}(p)(\phi^n \circ \gamma)'(0)$$

This formula holds under certain differentiability assumptions which we shall discuss in section 2.2.8. It is more familiar in an abbreviated form in which γ is suppressed. For example, if P is a plane, say with a polar co-ordinate system (r, θ), then a path is usually called a curve in parametric form and described as $(r(t), \theta(t))$, rather than $(r(\gamma(t)), \theta(\gamma(t)))$ whereupon the chain rule is expressed

$$f' = \frac{\partial f}{\partial r} r' + \frac{\partial f}{\partial \theta} \theta'$$

The two versions of the chain rule therefore correspond by replacing f, r, and θ in the expressions f', r' and θ' with $f \circ \gamma$, $r \circ \gamma$ and $\theta \circ \gamma$. Our formulation of the chain rule simply makes it explicit that when a function is regarded as a function of t, it is by evaluating it at the location $\gamma(t)$ of some given moving point or variation.

The chain rule implies the remarkable fact that the rate of change of a function under a variation γ depends upon the variation only through the rates of change $(\phi^1 \circ \gamma)'(0), \ldots, (\phi^n \circ \gamma)'(0)$ of the co-ordinate functions. In other words, if two variations γ_1 and γ_2 produce the same rates of change in each of the co-ordinate functions $\phi^1, \ldots, \phi^n$ then they will produce the same rate of change in every other function for which the chain rule holds. It follows that the variations through any fixed point p can be clumped together into classes by the criterion that two variations fall into the same class if the rate of change of each co-ordinate function is the same along both.

The classification of variations according to the rates of change of co-ordinate functions can be regarded as embodying a generalisation of the notion of velocity. Indeed, if the set P is a plane and the co-ordinate system is a Cartesian system (x, y) then, where the chain rule holds, the rate of change of a function under a variation γ depends only on $(x \circ \gamma)'(0)$ and $(y \circ \gamma)'(0)$. In classical notation these would be described simply as $x'(0)$ and $y'(0)$. They are traditionally interpreted as the x and y components of the velocity at $p = \gamma(0)$ of the moving point described by γ. If two variations through a point p induce the same rates of change in x and y then we say they have the same velocity. In a polar co-ordinate system (r, θ) the numbers $(r \circ \gamma)'(0)$ and $(\theta \circ \gamma)'(0)$ are called the radial and angular components of the velocity of the variation. Although these numbers differ from the Cartesian case, they are still taken to parametrise the same property of variations, namely their velocity through p.

Following this example, we shall regard the rates of change of the co-ordinate functions of a co-ordinate system as reflecting some concept of generalised velocity of variations through a point. For an abstract set, such as a set of probability distributions, or thermodynamic states of a mass of gas, or the events which constitute the cosmos, there can be no concept of the velocity of a variation through an element p in the sense of its direction and speed. The velocity of a variation in the generalised sense is not *a priori* a combination of concepts such as speed and direction, and only reduces to this kind of description for special kinds of sets like a plane which admit such concepts. A generalised velocity can really be understood only as a class v of variations, which are considered to share the same generalised velocity, as determined by the fact that the rates of change of the co-ordinates functions are constant over this class. We will say that a generalised velocity v is represented by any one of its constituent variations γ, or that γ has velocity v when γ belongs to the class v. In practice, as in the Cartesian case, one distinguishes generalised velocities by the values $(\phi^1 \circ \gamma)'(0), \ldots, (\phi^n \circ \gamma)'(0)$. Given a variation γ through p we shall denote its velocity by $\gamma'(0)$. This isn't meant to say that its velocity is some object derived by operations upon γ itself. We use the notation $\gamma'(0)$ merely to denote the larger collection of variations to which γ belongs, all of which produce the same rates of change for the co-ordinate functions.

In addition to determining a notion of generalised velocity for variations through a point, co-ordinate systems also determine an addition and scalar multiplication of these velocities, by means of componentwise addition and scalar multiplication of the parameters $(\phi^1 \circ \gamma)'(0), \ldots, (\phi^n \circ \gamma)'(0)$. Specifically, the componentwise addition and scalar multiplication of parameters are simply taken to correspond to an addition and scalar multiplication of generalised velocities themselves. The set of all velocities at a point p thereby acquires a vector space structure. In order to emphasise this vector space structure the set of all velocities at p is called the *tangent space* of the set P at the element p and we denote it T_pP. Its elements, the velocities at p, are then called *tangent vectors.*

It is important to remember that tangent vectors, as we have defined them here, are simply velocities in the generalised sense, and are not in the first instance interpretable as lines or arrows tangent to a surface, as happens in geometry or as we interpreted them in Chapter 1. That use of tangent vector in section 1.5.1 and the examples which we give below show how such geometric tangent vectors correspond to velocities in the usual sense, and so justify adopting the name tangent vector for generalised velocities. In section 3.1 we discuss the geometric realisation of tangent vectors in more detail. However, the purpose of our discussion so far has been to show that, even in the most abstract setting where geometrical interpretation seems quite inappropriate, a parametrisation of a set can be treated as a co-ordinate system, and the chain rule of multivariable calculus introduces a generalisation of the concept of velocity at each point, together with linear operations on this set of velocities. The interpretations are quite indifferent to geometry, although geometrical examples can be considered special cases from which the terminology arises.

Example 2.2.4 *Tangent vectors in* $\mathbf{R}^n$. Let $P = \mathbf{R}^n$. For most sets there is no distinguished choice of co-ordinate system. Even in the plane, where Cartesian co-ordinate systems are distinguished by Euclidean geometry, there are an infinity of such co-ordinate systems varying by their choice of origin and orientation of axes. On $\mathbf{R}^n$ however, the components of its elements provide a distinguished co-ordinate system. Let x^i be the function on $\mathbf{R}^n$ whose value on any element a is its ith component. Thus, instead of reading $x^i(a)$ in the usual way as '$x - i - of - a$' we should replace 'x' by 'component' and read it as 'component $- i - of - a$'.

Usually one writes a^i for the value $x^i(a)$, i.e. for the ith component of a. Likewise for a variation γ in $\mathbf{R}^n$ one usually writes γ^i instead of $x^i \circ \gamma$. Because of this distinguished co-ordinate system $\mathbf{R}^n$ provides the one instance where we usually do identify the tangent vector of a variation γ through a point with the list of numbers $(x^1 \circ \gamma)'(0), \ldots, (x^n \circ \gamma)'(0)$, which is to say the list $(\gamma^1)'(0), \ldots, (\gamma^n)'(0)$. In particular, each tangent vector through a point a in $\mathbf{R}^n$ is identified with an element of $\mathbf{R}^n$. Indeed, for variations γ through a point a the map

$$\gamma \mapsto ((\gamma^1)'(0), \ldots, (\gamma^n)'(0))$$

defines a linear isomorphism

$$T_a\mathbf{R}^n \to \mathbf{R}^n$$

Example 2.2.5 *Tangent vectors in a vector space.* In a vector space V a choice of basis $v_1, \ldots, v_n$ provides a linear co-ordinate system $\phi^1, \ldots, \phi^n$, where the values $\phi^1(v), \ldots, \phi^n(v)$ are the coefficients in the expansion of v in terms of $v_1, \ldots, v_n$. Every co-ordinate system which is linear, i.e. such that the co-ordinate functions $\phi^1, \ldots, \phi^n$ are linear, arises in this way. A basis and the linear co-ordinate system which it defines are said to be dual to each other, and determine each other by the relationship

$$\phi^i(v_j) = \delta^i_j$$

The velocity of variations γ are parametrised by the lists of numbers $\big((\phi^1 \circ \gamma)'(0), \ldots, \phi^n \circ \gamma)'(0)\big)$, and these vary with the choice of co-ordinates. However, for linear co-ordinates it is easy to show that the vector

$$(\phi^1 \circ \gamma)'(0)v_1 + \ldots + (\phi^n \circ \gamma)'(0)v_n$$

is independent of the choice of dual pair $\phi^1, \ldots, \phi^n$ and $v_1, \ldots, v_n$. We call this vector $\gamma'(0)$, and we note that it is effectively defined by the formula

$$\phi^i(\gamma'(0)) = (\phi^i \circ \gamma)'(0) \tag{2.2.1}$$

By this construction each velocity at a vector p in V is itself identified with a vector in V so the map $\gamma \mapsto \gamma'(0)$ defines a map $T_pV \to V$. Because of equation 2.2.1 it is obviously linear.

The identification between velocities at a point p in V and V itself can be made more directly. Suppose we say that a V-valued function f is continuous if $\phi \circ f$ is continuous for every linear function ϕ. Then we can say that for any V-valued function g,

$$\lim_{t \to 0} g$$

is the vector w which, taken as the value of g at 0, makes g continuous at 0. Then we can define

$$\gamma'(0) = \lim_{t \to 0} \frac{\gamma(t) - \gamma(0)}{t}$$

The map $\gamma \mapsto \gamma'(0)$ from variations through p to V gives a welldefined map $T_pV \to V$ which is obviously linear. Moreover, for any v in V the path $\gamma_v(t) = p+tv$ clearly has $\gamma'(0) = v$, so that $v \mapsto \gamma_v$ inverts this map.

Once again, as in the first example, the elements of V play a double role, as both points of the space and as velocities. In its role as a velocity an element v in V is often depicted as an arrow based at p, and is called a tangent vector. The uniform variation $p+tv$ is considered an archetypal variation with velocity v, and shows once again how geometric tangent vectors determine tangent vectors in the sense of generalised velocities.

Example 2.2.6 *Tangent vectors in an affine space.* If P is an affine space, modelled on a vector space V then we can use the same approach as for vector spaces to identify velocities with translation vectors. Given any point p in P a variation γ through p determines a variation $\bar{\gamma}$ in V through 0 by defining

$$\gamma(t) \;=\; p \;+\; \bar{\gamma}(t)$$

Remember that $p+v$ means p translated through v, and note that $\bar{\gamma}(0) = 0$. The association $\gamma \mapsto \bar{\gamma}'(0)$ defines a map $T_pP \to V$. Once again, it is easy to show that the map is linear, and that $v \mapsto p+tv$ inverts it.

What the inverse tells us is that any translation vector v determines a unique velocity at p, which is the velocity containing the uniform variation $p+tv$.

Note that we don't really need to restrict ourselves to translation only from the point p through which the variation passes. If q is any other point then we can define $\tilde{\gamma}$ by

$$\gamma(t) \;=\; q \;+\; \tilde{\gamma}(t)$$

There is a unique translation v such that $q = p+v$, so that $\bar{\gamma}(t) = v + \tilde{\gamma}(t)$ and the derivatives of both $\bar{\gamma}$ and $\tilde{\gamma}$ are the same.

Example 2.2.7 *Tangent vectors to the space of probability distributions.* If we let $P = \mathcal{M}$ then the fact that $\mathcal{M}$ has an infinite-dimensional vector space of translations makes little difference to the discussion in example 2.2.6. A variation through a measure ν in $\mathcal{M}$ is just a one-parameter family of measures $\nu(t)$ with $\nu(0) = \nu$. As a parametric family with one parameter t, the notions of log-likelihood and score apply to it, and realise the the isomorphism

$T_\nu \mathcal{M} \to R_\Omega$ from the tangent space of the affine space $\mathcal{M}$ to its vector space of translations, described in example 2.2.6 above.

Indeed, if $p(t)$ is the density of $\nu(t)$ with respect to ν then according to the log-likelihood affine structure $\bar{\nu}(t) = \log(p(t))$; it should be denoted $\ell(\nu(t))$ and called the log-likelihood of $\nu(t)$ relative to ν. The definition we have given for $\bar{\nu}'(0)$ is a little more problematic, since the notion of continuity which we have proposed is not really suitable for infinite-dimensional vector spaces like R_Ω. However, we shall restrict ourselves to situations where $p(t)$ is itself a function on the sample space (rather than just an equivalence class to within measure zero), so that we can express it as $p(t, x)$ where x refers to a point in the sample space. Then the t derivative of $\log(p(t, x))$ can be taken as its partial derivative with respect to t. This is the score of the parametric family $\nu(t)$. We regard two variations through ν as having the same velocity if their scores, defined in this sense, are the same. Taking the score therefore defines the isomorphism $T_\nu \mathcal{M} \to R_\Omega$ from the tangent space of the affine space $\mathcal{M}$ to its vector space of translations.

In terms of the tangent space of $\mathcal{M}$, the tangent space to the space of all probability distributions $\mathcal{P}$ can be understood in two different ways. First, considering $\mathcal{P}$ as a subset of $\mathcal{M}$, we can identify $T_\nu \mathcal{P}$ as a subspace of $T_\nu \mathcal{M}$. If $\nu(t)$ is a variation of probability measures through a probability measure ν then, as above, $\bar{\nu}(t) = \log(p(t))$, where $p(t)$ is the density of $\nu(t)$ with respect to ν. The fact that $\nu(t)$ is a probability measure implies that

$$\int_\Omega \exp(\bar{\nu}(t)) d\nu = 1$$

for all t. Differentiating at $t = 0$ gives the equation

$$E_\nu(\bar{\nu}'(0)) = 0$$

Hence those random variables f in R_Ω which represent tangent vectors to ν within the space of probability measures $\mathcal{P}$ are the ones satisfying the condition $E_\nu(f) = 0$. Conversely if $E_\nu(f) = 0$ then the variation $\nu(t) = \nu + tf$, that is, the family defined by

$$d\nu(t) = \exp(tf) d\nu$$

and dividing by its total mass gives a family of probability measures through ν whose velocity is represented by f.

On the other hand $\mathcal{P}$ is an affine space in its own right, regarded as finite positive measures up to scale. Its tangent spaces can be identified with its own vector space of translations, namely $\mathbf{R}_\Omega/\mathbf{R}.1$ or functions identified up to the addition of a constant. Notice that by subtracting the expected value of a function f from itself we produce a function within its equivalence class with expectation zero, and that each equivalence class contains only one such function. Hence the map

$$\{f \mid E_\nu(f) = 0\} \subset R_\Omega \to R_\Omega/\mathbf{R}.1$$

which sends f to $f + \mathbf{R}.1$ is an isomorphism, and identifies the two versions of the tangent spaces of $\mathcal{P}$.

Example 2.2.8 *Tangent vectors to a parametrised family.* The score of a parametric family P of probability distributions is actually a list of tangent vectors to the family considered as a manifold inside $\mathcal{M}$. As with the set of all probability distributions, any variation $\nu(t)$ in P is also a variation in $\mathcal{M}$, and as we described in example 2.2.7,we have

$$\bar{\nu}(t) \; = \ell(\nu(t)) \; = \log(p(t))$$

where $p(t)$ is the density of $\nu(t)$ with respect to some base measure. Given that P is parametric we can form such variations by varying the parameters of the family, and in particular we consider

$$\bar{\nu}(t) \; = \ell(\nu(t)) \; = \log(p(\theta^1, \dots \theta^i + t, \dots \theta^n))$$

so that $\bar{\nu}'(0)$ is just the ith partial derivative of the log-likelihood, i.e. the ith component of the score.

2.2.4 *Co-ordinate independence*

The concept of tangent vectors, as generalised velocities, and the addition and scalar multiplication of the tangent vectors located at any particular point, are made using a co-ordinate system. It may be that different co-ordinate systems give different definitions of tangent vector or of the linear operations upon tangent vectors. For example, in the plane we will say that two variations γ_1 and

γ_2 of a point have the same tangent vector if, for some Cartesian co-ordinate system (x, y) we find that $(x \circ \gamma_1)'(0) = (x \circ \gamma_2)'(0)$ and $(y \circ \gamma_1)'(0) = (y \circ \gamma_2)'(0)$. However, it may be that for some other co-ordinate system (ϕ^1, ϕ^2) we find that either $(\phi^1 \circ \gamma_1)'(0)$ is not equal to $(\phi^1 \circ \gamma_2)'(0)$ or $(\phi^2 \circ \gamma_1)'(0)$ is not equal to $(\phi^2 \circ \gamma_2)'(0)$. Hence, according to the co-ordinate system (ϕ^1, ϕ^2), the variations γ_1 and γ_2 don't have the same velocity. Indeed, since there are many different Cartesian co-ordinate systems, with different origins, orientations and scales of length, one can legitimately ask whether all of these give the same definition of the generalised velocity of variations of points and the linear operations on them, i.e. give the same definition of tangent spaces at each point.

Let us now observe that the chain rule not only allows the definition of tangent space, but solves the problem of consistency of its definition as well. The concept of generalised velocity, as the determinant of the rates of change of functions, applies only to those functions satisfying the chain rule. We can show that co-ordinate systems whose co-ordinate functions satisfy the chain rule with respect to each other, all give the same definition of the tangent spaces at each point, that is, the same definition of generalised velocity, and of the linear operations on velocities. For suppose that the co-ordinate functions of the co-ordinate system $\psi^1, \ldots, \psi^n$ satisfy the chain rule with respect to the co-ordinate system $\phi^1, \ldots, \phi^n$. Then the rates of change of each of the functions $\psi^i, i = 1, \ldots, n$ under a variation γ is determined by the rates of change of $\phi^1, \ldots, \phi^n$ under γ. If these latter are equal for two different variations, then the rates of change of each of the functions $\psi^1, \ldots, \psi^n$ must also be the same along these two variations. Hence two variations which are declared to have the same velocity by $\phi^1, \ldots, \phi^n$, will also be declared to have the same velocity by $\psi^1, \ldots, \psi^n$. Conversely, if each of the functions $\phi^i, i = 1, \ldots, n$ satisfies the chain rule with respect to the co-ordinate system $\psi^1, \ldots, \psi^n$, then any two paths which are declared to have the same velocity by $\psi^1, \ldots, \psi^n$, will also be declared to have the same velocity by $\phi^1, \ldots, \phi^n$. The chain rule relations between the co-ordinate systems can be expressed as

$$(\psi^i \circ \gamma)'(0) = \frac{\partial \psi^i}{\partial \phi^1}(p)(\phi^1 \circ \gamma)'(0) + \ldots + \frac{\partial \psi^i}{\partial \phi^n}(p)(\phi^n \circ \gamma)'(0)$$

and

$$(\phi^i \circ \gamma)'(0) = \frac{\partial \phi^i}{\partial \psi^1}(p)(\psi^1 \circ \gamma)'(0) + \ldots + \frac{\partial \phi^i}{\partial \psi^n}(p)(\psi^n \circ \gamma)'(0)$$

It follows at once that the addition and scalar multiplication of velocities defined by the componentwise operations on

$$(\phi^1 \circ \gamma)'(0), \ldots, (\phi^n \circ \gamma)'(0)$$

is the same as that defined by

$$(\psi^1 \circ \gamma)'(0), \ldots, (\psi^n \circ \gamma)'(0)$$

Because of these results we shall say that two co-ordinate systems $\phi^1, \ldots, \phi^n$ and $\psi^1, \ldots, \psi^n$ are *differentiably related* if each function from one satisfies the chain rule with respect to the other. We can then say that differentiably related co-ordinate systems define the same tangent spaces at each point. Moreover, it is easy to see that differentiable relatedness is an equivalence relation on the set of all co-ordinate systems. Hence, by remaining within any one equivalence class of co-ordinate systems, we can be sure that the definition of the tangent spaces is independent of the co-ordinate system chosen from this class. For example, it is a simple matter to show that all Cartesian co-ordinate systems are differentiably related, and so they all determine the same tangent spaces at each point of the plane.

In 2.2.8 we show that the chain rule holds if the co-ordinate expression of a function is differentiable on $\mathbf{R}^n$ in the sense of being linearisable. If each of the partial derivatives of the co-ordinate expression exist and are continuous then such linearisability is guaranteed. This provides a simple test for differentiable relatedness. For convenience we work with *smooth* functions, that is those which are differentiable not merely to first but to all orders. We say that two co-ordinate systems ϕ and ψ are *smoothly related* if the co-ordinate expressions of the functions ψ are smooth when expressed using the co-ordinates ϕ, and vice versa.

Example 2.2.9 *Affine spaces.* Let P be an affine space. Let θ and χ be affine co-ordinates; then there exists a matrix X^i_j and a vector v^i such that

$$\theta^i = \sum_j X^i_j \chi^j + v^i$$

Thus we see that

$$\frac{\partial\theta^i}{\partial\chi^j} = X^i_j$$

so these two co-ordinate systems are smoothly related.

Example 2.2.10 *The normal family.* Consider the space $\mathcal{N}$ of all normal distributions. Recall from Chapter 1 that the canonical co-ordinates

$$\theta = (\theta^1, \theta^2): \mathcal{N} \to \mathbf{R}^2$$

are related to the (μ, σ) co-ordinates by

$$\theta^1 = \frac{-1}{2\sigma^2} \quad \text{and} \quad \theta^2 = \frac{\mu}{\sigma^2}$$

Notice that the map

$$\theta = (\theta^1, \theta^2): \mathcal{N} \to \mathbf{R}^2$$

which sends p to $\theta(p) = (\theta^1(p), \theta^2(p))$ has image the set

$$\{(\theta^1, \theta^2) | \theta^1 < 0\}$$

and that it is one-to-one. It follows that it defines co-ordinates on $\mathcal{N}$. Clearly θ^1 and θ^2 are smoothly related to the co-ordinates μ and σ.

2.2.5 *Differentials and 1-forms*

It is the essence of the concept of tangent vectors to variations that it determines the rates of change of functions, so that the rate of change of any function f is to be considered a function of tangent vectors only, and not the fine detail of the individual variations themselves. We denote this function of tangent vectors by df and call it the differential of f. Thus df is defined on tangent vectors v by

$$df(v) = (f \circ \gamma)'(0)$$

where γ is any variation with tangent vector v, i.e. any variation in the collection corresponding to v. Notice that df does not mean 'an infinitesimal increment in f'. The definition of $df(v)$ also explains the most direct way to calculate it, namely to choose a path γ representing v and to compute the rate of change of f along it. On the other hand one usually describes tangent vectors by specifying what the values of $(\phi^1 \circ \gamma)'(0), \ldots, (\phi^n \circ \gamma)'(0)$ are to be, which by definition are the values $d\phi^1(v), \ldots, d\phi^n(v)$. In this case one can calculate $df(v)$ using the chain rule which, from the definition of differential, assumes the form

$$df(v) = \frac{\partial f}{\partial \phi^1}(p)d\phi^1(v) + \cdots + \frac{\partial f}{\partial \phi^n}(p)d\phi^n(v)$$

The chain rule thus becomes an identity between functions of tangent vector, and if we leave out the argument of the functions, as is usual when expressing an identity between functions, we obtain the formula for the so-called total differential

$$df = \frac{\partial f}{\partial \phi^1}d\phi^1 + \cdots + \frac{\partial f}{\partial \phi^n}d\phi^n$$

Our interpretation of the total differential stands in contrast with its interpretation in the calculus of several variables. In calculus on manifolds each side of the formula for the total differential is defined independently as a function of tangent vectors, and the formula itself is an identity between these functions. In several-variable calculus the formula often serves as a kind of definition of df. Indeed f is regarded as a function of the variables $\phi^1, \ldots, \phi^n$ and df as an infinitesimal increment in its values, rather than a function of tangent vectors. $\phi^1, \ldots, \phi^n$, in their role as independent variables, are considered a different species from f. Since as independepent variables $\phi^1, \ldots, \phi^n$ are not considered functions of anything else, the idea of df is not meant to apply to them, and they are more often regarded as infinitesimal numbers of some kind. In calculus on manifolds both f and $\phi^1, \ldots, \phi^n$ are interpreted as species of the same kind, namely functions on a given set, and in particular the definition of the differential of a function applies equally well to the co-ordinate functions. One should always translate the letter d in this context as 'the rate of change of', rather than as 'an infinitesimal increment of' so that df translates as 'the rate of change of f', and $df(v)$ as 'the rate of change of f along (the tangent vector) v'.

If v_1 and v_2 belong to T_pP, i.e. they are tangent vectors at the point p, then the definition of $v_1 + v_2$, is designed so that the differentials of the co-ordinate functions $\phi^1, \dots \phi^n$ are linear. The chain rule, or equivalently the formula for the total differential, then implies that df is linear on T_pP for every function f for which the chain rule holds. Hence

$$df(v_1 + v_2) = df(v_1) + df(v_2)$$

and for each $v \in T_pP$ and real number λ we have

$$df(\lambda v) = \lambda df(v)$$

In general let us denote the set of all tangent vectors at all points of P by TP. It is called the *tangent bundle* of P. Each element v of TP is a tangent vector or tangent vector at some point p, which we call the position or location of v. We sometimes denote it by $\pi(v)$ and call π the position map, since it defines a map $\pi: TP \to P$ whose value on any tangent vector v is its position. Functions on TP are simply functions of tangent vectors, and the restriction of any function ω on the tangent bundle TP to T_pP, the tangent vectors located at a specific point p, is denoted by ω_p. The only examples of functions on the tangent bundle, i.e. functions of tangent vectors, that we have introduced so far are the functions df. They have the special property that each restriction df_p is linear as a function on the vector space T_pP. General functions ω on TP with this property are called 1-forms, and are known classically as covariant tensors of rank 1.

Given the co-ordinate system $\phi^1, \dots, \phi^n$ we have through each point p in the set P the co-ordinate variations $\gamma_1, \dots, \gamma_n$, which are used to define the partial derivatives of functions at p. We denote the tangent vector of each of these variations through p by

$$\frac{\partial}{\partial \phi^1}(p), \dots, \frac{\partial}{\partial \phi^n}(p)$$

We use this notation because these tangent vectors are the ones which define the partial derivatives of functions, and it is an exercise in definitions that

$$\frac{\partial f}{\partial \phi^j}(p) = df(\frac{\partial}{\partial \phi^j}(p)) \tag{2.2.2}$$

In particular, since the co-ordinate variation γ_j is designed so that ϕ^i is constant except when $i = j$, where it has rate of change 1, it follows that

$$d\phi^i(\frac{\partial}{\partial\phi^j}(p)) = \delta^i_j$$

where δ^i_j is the Kronecker delta symbol which has the value 1 if $i = j$ and 0 otherwise.

2.2.6 *Vector fields*

For each $p \in P$, the tangent vector

$$\frac{\partial}{\partial\phi^j}(p)$$

of a co-ordinate variation is a tangent vector whose position is p. Hence we can regard it as a map to be evaluated on points $p \in P$. It is a map $P \to TP$, and the condition 'position of $(\partial/\partial\phi^j)(p)$ is p' can be written

$$\pi(\frac{\partial}{\partial\phi^j}(p)) = p$$

where π is the position map. General maps $X : P \to TP$ such that $\pi(X(p)) = p$, are called vector fields, and in particular we call

$$\frac{\partial}{\partial\phi^1}, \ldots, \frac{\partial}{\partial\phi^n}$$

the co-ordinate vector fields.

Classically a vector field on the plane is described as a field of arrows, meaning an arrow based at each point of the plane. However, as was pointed out in example 2.2.6, arrows can be identified with tangent vectors, any arrow based at a point p determining the uniform variation through p, which passes through p at time zero and translates a moving point through a copy of the arrow in each unit of time.

Let v be a tangent vector in the plane located at a point p. It corresponds to an arrow based at p, describing a uniform motion which goes from p to the tip of the arrow in a unit time. One can show that the rate of change of the x and y co-ordinates of a Cartesian co-ordinate system are therefore given by the difference between their values at the tip and at the base p of the arrow. These are just the directed distances which are the projections of v along the x and y axes. Thus $dx(v)$ and $dy(v)$ are equal to the 'horizontal' and 'vertical' components of the arrow v.

Vector fields on the plane are commonly expressed in the form

$$a\mathbf{i} + b\mathbf{j}$$

where a and b are functions, $\mathbf{i}$ is the vector field consisting of horizontal arrows of unit length and $\mathbf{j}$ is the vector field of vertical arrows of unit length. These are actually the co-ordinate vector fields $\partial/\partial x$ and $\partial/\partial y$ of a Cartesian co-ordinate system with horizontal x-axis and vertical y-axis. Indeed, the uniform motion through any point determined by $\mathbf{i}$ is a horizontal motion with unit speed in the positive x direction, and this is exactly the x co-ordinate path. Similarly $\mathbf{j}$ determines the y co-ordinate path through any point. In general, co-ordinate vector fields in the plane can be determined from the relations

$$d\phi^i\left(\frac{\partial}{\partial\phi^j}\right) = \delta^i_j$$

which say that $\partial/\partial\phi^j$ should induce rate of change zero in all of the co-ordinate functions except ϕ^j, whose rate of change should be 1. Thus knowing, for example, that $dx(v)$ and $dy(v)$ correspond to the horizontal and vertical projections of an arrow v, the condition $dy(\partial/\partial x) = 0$ says that $\partial/\partial x$ must be a horizontal arrow , while $dx(\partial/\partial x) = 1$ says that it must have unit length. Similarly we can infer that $\partial/\partial y$ is a vertical arrow with unit length. For a polar co-ordinate system (r, θ) and any arrow v we have that $dr(v)$ is the rate of change of r, the distance of the moving point from the co-ordinate origin, under the uniform motion determined by v, while $d\theta(v)$ is the rate of change of angle. Thus $d\theta(\partial/\partial r) = 0$ says that $(\partial/\partial r)(p)$ points radially, i.e. along the line from the origin to p, while $dr(\partial/\partial r) = 1$ implies that its length is 1, since r measures distances. On the other hand, $dr(\partial/\partial\theta) = 0$ implies that the arrow $(\partial/\partial\theta)(p)$ must be tangent to a line of constant r, i.e. a circle centred at the origin and passing through p. Moreover, since θ does not measure distances, $(\partial/\partial\theta)(p)$ doesn't have unit length. It is $r\theta$ which measures length around circles centred on the origin, and since $d\theta(\partial/\partial\theta) = 1$, we have $rd\theta((\partial/\partial\theta)(p)) = r(p)$, showing that the length of the arrow $(\partial/\partial\theta)(p)$ equals $r(p)$.

2.2.7 *Co-ordinate calculations*

We pointed out in the section above that a vector field on the plane is often described in the form $a\mathbf{i} + b\mathbf{j}$, where $\mathbf{i}$ and $\mathbf{j}$ are the horizontal and vertical unit length vector fields, and a and b are functions. From classical vector analysis the values $a(p)$ and $b(p)$ at any point p are known to be the horizontal and vertical projections of the arrow at p determined by this vector field. We have pointed out that $\mathbf{i}$ and $\mathbf{j}$ are in fact the co-ordinate vector fields $\partial/\partial x$ and $\partial/\partial y$ form a Cartesian co-ordinate system with horizontal x axis. We also showed that the horizontal and vertical projections of a tangent vector v are given by $dx(v)$ and $dy(v)$. Hence we obtain the expansion

$$v = dx(v)\frac{\partial}{\partial x} + dy(v)\frac{\partial}{\partial y}$$

This is a typical expansion of a tangent vector in terms of a co-ordinate system, and we want to show that, rather than depending on all of the intermediate geometrical interpretations, such expansions can be derived directly, and hold for any co-ordinate system on any abstract set.

Keeping to the example of the plane, suppose we assume that a tangent vector $v \in T_pP$ has an expansion

$$v = a\frac{\partial}{\partial x}(p) + b\frac{\partial}{\partial y}(p) \qquad (2.2.3)$$

Applying dx to both sides of (2.2.3), and using the linearity of dx, one has

$$dx(v) = adx(\frac{\partial}{\partial x}(p)) + bdx(\frac{\partial}{\partial y}(p)) = a.1 + b.0 = a$$

Similarly one derives that $dy(v) = b$. Hence if any tangent vector has an expansion of the form (2.2.3), then that expansion must be the co-ordinate expansion. Let us now show that any tangent vector must have such an expansion. We consider

$$v - dx(v)\frac{\partial}{\partial x}(p) - dy(v)\frac{\partial}{\partial y}(p)$$

Applying both dx and dy to the tangent vector gives zero. But this identifies it as the zero tangent vector. It follows that $v = dx(v)\partial/\partial x + dy(v)\partial/\partial y$, i.e. the expansion holds.

This same argument applies for any co-ordinate system on any abstract set. For suppose that $\phi^1, \ldots, \phi^n$ is a co-ordinate system on a set P. For any tangent vector $v \in T_pP$ consider $v - d\phi^1(v)(\partial/\partial\phi^1)(p) - \ldots - d\phi^n(v)(\partial/\partial\phi^n)(p)$. Applying $d\phi^j$ to this tangent vector gives

$$d\phi^j(v) - d\phi^1(v)d\phi^j(\frac{\partial}{\partial\phi^1}(p)) - \ldots - d\phi^n(v)d\phi^j(\frac{\partial}{\partial\phi^n}(p))$$
$$= d\phi^j(v) - d\phi^j(v) = 0$$

for $j = 1, \ldots, n$. Hence this tangent vector is the zero vector, and

$$v = d\phi^1(v)\frac{\partial}{\partial\phi^1}(p) + \ldots + d\phi^n(v)\frac{\partial}{\partial\phi^n}(p)$$

The fact that every tangent vector has a unique expansion in the form says that for any co-ordinate system $\phi^1, \ldots, \phi^n$ on a set P, the values $(\partial/\partial\phi^1)(p), \ldots, (\partial/\partial\phi^n)(p)$ of the co-ordinate vector fields at p form a basis for the tangent space T_pP at p.

If X is any vector field on P, then evaluating it at a point p and using the co-ordinate expansion above gives

$$X(p) = d\phi^1\left(X(p)\right)\frac{\partial}{\partial\phi^1}(p) + \ldots + d\phi^n\left(X(p)\right)\frac{\partial}{\partial\phi^n}(p)$$

In other words, any vector field has a unique expansion of the form

$$X = a_1\frac{\partial}{\partial\phi^1} + \ldots + a_n\frac{\partial}{\partial\phi^n}$$

where the co-efficients a_i are the functions

$$a_i(p) = d\phi^i\left(X(p)\right)$$

Now suppose that α is any linear function on T_pP. Applying it to the co-ordinate expansion, and using the linearity of α, gives

$$\alpha(v) = d\phi^1(v)\alpha\left(\frac{\partial}{\partial\phi^1}(p)\right) + \ldots + d\phi^n(v)\alpha\left(\frac{\partial}{\partial\phi^n}(p)\right)$$

Since this holds for every $v \in T_pP$ we have the identity of functions

$$\alpha = \alpha\left(\frac{\partial}{\partial\phi^1}(p)\right)d\phi^1 + \ldots + \alpha\left(\frac{\partial}{\partial\phi^n}(p)\right)d\phi^n$$

The fact that every linear function on T_pP has a unique expansion of the form says that for any co-ordinate system $\phi^1, \ldots, \phi^n$ on a set P, the restrictions $d\phi^1_p, \ldots, d\phi^n_p$ of co-ordinate 1-forms to T_pP form a basis of the vector space dual to T_pP, i.e. the space of all linear functions on T_pP.

If ω is a 1-form on P then we may apply the expansion to its restriction to T_pP, obtaining

$$\omega_p = \omega_p \left(\frac{\partial}{\partial \phi^1}(p) \right) d\phi^1 + \ldots + \omega_p \left(\frac{\partial}{\partial \phi^n}(p) \right) d\phi^n$$

Since this is an identity in p, it follows that any 1-form ω has a unique expansion of the form

$$\omega = a_1 d\phi^1 + \ldots + a_n d\phi^n \tag{2.2.4}$$

where $a_1, \ldots, a_n$ are the functions

$$a_j(p) = \omega_p \left(\frac{\partial}{\partial \phi^j}(p) \right) \tag{2.2.5}$$

If the formulas (2.2.2) and (2.2.5) are applied to the 1-form df then one recovers the formula for the total differential.

These co-ordinate formulas for vector fields and 1-forms are often used to express the co-ordinate vector fields and 1-forms of one co-ordinate system in terms of the co-ordinate vector fields and 1-forms of another. Let us calculate some examples:

Example 2.2.11 *Polar co-ordinates.* If a polar co-ordinate system (r, θ) on the plane has the same origin as a Cartesian co-ordinate system (x, y), and the positive x-axis as its defining ray, then the co-ordinate functions are related by

$$x = r\cos\theta \quad \text{and} \quad y = r\sin\theta$$

These are, of course, identities between functions on the plane, so that for example the first identity means

$$x(p) = r(p)\cos\theta(p)$$

for each point p in the plane.

Using the formula for the total differential we obtain

$$dx = \cos\theta dr - r\sin\theta d\theta$$
$$dy = \sin\theta dr + r\cos\theta d\theta$$

Treating these as equations for dr and $d\theta$, we can solve them to express the co-ordinate differentials dr and $d\theta$ in terms of dx and dy:

$$dr = \cos\theta dx + \sin\theta dy$$
$$d\theta = \frac{-\sin\theta}{r}dx + \frac{\cos\theta}{r}dy$$

For the co-ordinate vector fields equation (2.2.2) gives

$$\frac{\partial}{\partial r} = a\frac{\partial}{\partial x} + b\frac{\partial}{\partial y} \tag{2.2.5}$$

for some functions a and b. Applying dx to both sides of (2.2.5) gives

$$dx(\frac{\partial}{\partial r}) = adx(\frac{\partial}{\partial x}) + bdx(\frac{\partial}{\partial y}) = a$$

so that $a = \partial x/\partial r = \cos\theta$. Similarly, applying dy to both sides of (2.2.5) gives $b = \sin\theta$. Hence we obtain

$$\frac{\partial}{\partial r} = \cos\theta\frac{\partial}{\partial x} + \sin\theta\frac{\partial}{\partial y} \tag{2.2.6}$$

The same arguments in respect of $\partial/\partial\theta$ gives

$$\frac{\partial}{\partial\theta} = -r\sin\theta\frac{\partial}{\partial x} + r\cos\theta\frac{\partial}{\partial y} \tag{2.2.7}$$

In classical vector analysis $\partial/\partial r$ is usually denoted $\mathbf{i}_r$ and (2.2.6) and (2.2.7) then become

$$\mathbf{i}_r = \cos\theta\mathbf{i} + \sin\theta\mathbf{j}$$
$$\mathbf{i}_\theta = -\sin\theta\mathbf{i} + \cos\theta\mathbf{j}$$

which are the formulas often used to picture the vector fields $\mathbf{i}_r$ and $\mathbf{i}_\theta$.

Example 2.2.12 *The normal family.* For the normal family and the (μ, σ) and canonical co-ordinates we have

$$\frac{\partial\theta^1}{\partial\mu} = 0 \quad \frac{\partial\theta^1}{\partial\sigma} = \frac{1}{\sigma^3}$$
$$\frac{\partial\theta^2}{\partial\mu} = \frac{1}{\sigma^2} \quad \frac{\partial\theta^2}{\partial\sigma} = \frac{-2\mu}{\sigma^3}$$

so that

$$d\theta^1 = \frac{1}{\sigma^3}d\sigma$$

and

$$d\theta^2 = \frac{1}{\sigma^2}d\mu - \frac{2\mu}{\sigma^3}d\sigma$$

2.2.8 *Differentiability and the chain rule*

This section is not necessary for the rest of the book. In it we make some remarks about the chain rule as we have formulated it in section 2.2.3. It is also the key to the co-ordinate independence of these concepts, as explained in section 2.2.4. We shall now give some details of the concept of differentiability which underlies the chain rule. Readers who are less interested in foundational issues, and who are happy to take the chain rule on faith, need not concern themselves with this section. Full details of this material can be found in any standard text on calculus on $\mathbf{R}^n$, for instance Fleming (1987).

For functions defined on $\mathbf{R}^n$ there is the concept of differentiability in terms of their being linearizable about a point. If a is a point in $\mathbf{R}^n$, and g is a function defined in a neighbourhood of a, then one seeks to expand $g(a + h) - g(a)$ into a part linear in h plus a remainder. That is, one seeks an expansion

$$g(a + h) - g(a) = L(h) + R(h) \tag{2.2.8}$$

where L is linear and R is 'higher order' in h. If g is an arithmetic expression in the components of a then 'order' can be interpreted as polynomial degree, linear being the same as first order, and then it is clear what such an expansion means and how to obtain it. More generally, however, one needs a broader definition of 'higher order' and a means to divide $g(a+h) - g(a)$ up in this way. This requires a measure of size for elements of $\mathbf{R}^n$. For example, if $h = (h_1, \ldots, h_n)$ then

$$||h|| = |h_1| + \ldots + |h_n|$$

is a suitable measure of the size of h. One then says that a function R of h is of higher order if

$$\lim_{h \to 0} \frac{|R(h)|}{||h||} = 0 \tag{2.2.9}$$

Other measures of size for h are possible, such as

$$||h|| = (h_1^2 + \ldots + h_n^2)^{\frac{1}{2}}$$

and more generally one can consider any norm on $\mathbf{R}^n$ as a measure of size. The important point is that the choice of norm makes no difference; the notion of higher order is the same no matter which norm is used, although this ceases to be the case if one wants to extend these notions from $\mathbf{R}^n$ to vector spaces of infinite dimension.

Having introduced the concept of a function $R(h)$ being of higher order in h, there remains the problem of making a decomposition of $g(a+h) - g(a)$ as in (2.2.8), or more generally of deciding whether such a decomposition can be made. If one chooses any linear function L, then one obtains a decomposition (2.2.8) by putting $R(h) = g(a+h) - g(a) - L(h)$, and one must decide if the resulting function R is of higher order in h or not. One can show that there is no more than one choice of L for which R is of higher order in h, and if such a choice exists one says that g is differentiable at a, and that L is its differential at a.

For functions g on $\mathbf{R}$ a linear function has the form $h \mapsto mh$, and the differentiability criterion (2.2.9) is that

$$\lim_{h \to 0} \frac{g(a+h) - g(a) - mh}{h} = 0$$

It is easy to see that this is the same as the usual criterion for differentiability of functions of a single variable, and that the coefficient m is the derivative $g'(a)$. Replacing $a + h$ by x, and the function $g(a + h)$ by $g(a) + mh$, one obtains the function

$g(a) + m(x - a)$ whose graph is the tangent to the graph of g at a. In higher dimensions, if x and a are points in $\mathbf{R}^n$ and L is a linear function then $g(a) + L(x - a)$ is a function whose graph in $\mathbf{R}^{n+1}$ is a hyperplane through the point $(a, g(a))$ on the graph of g. The 'higher order' condition is then a condition on the hyperplane, in particular on the way in which it make contact with the graph of g. Thus the definition of differentiability is the exact analogue in higher dimensions of the usual single-variable notion, that is a definition of tangency and of the equation of the tangent hyperplane in terms of first-order contact.

The definition of differentiability applies equally well to maps $g : \mathbf{R}^n \to \mathbf{R}^m$ as to functions on $\mathbf{R}^n$. One requires the choice of a linear map $L : \mathbf{R}^n \to \mathbf{R}^m$ such that $R(h) = g(a+h) - g(a) - L(h)$ satisfies (2.2.9) but with $|R(h)|$ is replaced by $||R(h)||$ for some norm on $\mathbf{R}^m$. One can then prove the chain rule.

Theorem *Let $g : \mathbf{R}^n \to \mathbf{R}^m$ be differentiable at a with differential L, and $f : \mathbf{R}^m \to \mathbf{R}^p$ be differentiable at $g(a)$ with differential M. Then $f \circ g$ is differentiable at a with differential $M \circ L$.*

If L is a linear function then it is differentiable at every point with differential L. This observation applies in particular to the component functions $x_1, \ldots, x_m$ on $\mathbf{R}^m$. If a map $g : \mathbf{R}^n \to \mathbf{R}^m$ is differentiable at a then by the Theorem so are its component functions $g_i = x_i \circ g$. A converse of this can be proved also. In particular this proves that a variation $\gamma : (-\epsilon, \epsilon) \to \mathbf{R}^n$ is differentiable at 0 if and only if each component function $\gamma_i = x_i \circ \gamma$ is so, and its differential, which we denote $\gamma'(0)$, is $(\gamma_1'(0), \ldots, \gamma_n'(0))$. Moreover, for any $v \in \mathbf{R}^n$ and point a the variation $a + tv$ is differentiable with differential v. It follows that if g is a map differentiable at a with differential L then $g(a + tv)$ is differentiable, as a function of a single variable, and has differential $L(v)$. This provides the main way of calculating values of the differential L. But since L is linear one only needs to know its values on some basis. If we use the standard basis $e_1, \ldots, e_n$ then the resulting values, as columns in $\mathbf{R}^m$, form a matrix which is the Jacobian matrix of g at a.

The Theorem above is in fact the chain rule for differentiable maps from $\mathbf{R}^n$ to $\mathbf{R}^m$. The chain rule stated in section 2.2.3 is a consequence of it, and in a sense just a reinterpretation of it. Given a co-ordinate system $\phi^1, \ldots, \phi^n$ on a set we say that a function f

on the set is differentiable relative to this co-ordinate system if the function $\bar{f}$ on $\mathbf{R}^n$ which expresses it as a function of these co-ordinates is itself differentiable as a function on $\mathbf{R}^n$. We say that a variation γ is differentiable if $\phi \circ \gamma = (\phi^1 \circ \gamma, \ldots, \phi^n \circ \gamma)$ is differentiable. Since $f \circ \gamma = \bar{f} \circ (\phi \circ \gamma)$ it follows from the Theorem above that $f \circ \gamma$ is differentiable if both f and γ are differentiable relative to the co-ordinate system. Moreover, if L is the differential of $\bar{f}$ then

$$\begin{aligned}(f \circ \gamma)'(0) &= L\left((\phi^1 \circ \gamma)'(0), \ldots, (\phi^n \circ \gamma)'(0)\right) \\ &= L(e_1)(\phi^1 \circ \gamma)'(0) + \ldots + L(e_n)(\phi^n \circ \gamma)'(0)\end{aligned}$$

where e_i is the vector with a 1 in the ith place and zeros elsewhere. However, $L(e_i)$ is just the usual ith partial derivative of $\bar{f}$ and can be readily identified with $(\partial f/\partial \phi^i)(\gamma(0))$. Making this identification,the formula above is precisely the chain rule as announced in section 2.2.3.

2.2.9 *Manifolds*

A local co-ordinate system on a set is one whose domain is not necessarily the whole set, and whose image in $\mathbf{R}^n$ is an open set. An open set in $\mathbf{R}^n$ is one which, for each of its points, contains not only the point but also some 'rectangular' set $(a_1, b_1) \times \ldots \times (a_n, b_n)$ centred about the point. In this sense it surrounds each of its points. From another perspective one can say that an open set has no edge or boundary points. The reason that one requires co-ordinate systems to have an open image is to ensure that boundaries and interiors of subsets are consistently represented by their co-ordinates. A good example is a polar co-ordinate system on the plane. If one wants the domain of a polar co-ordinate system to be the whole plane then $r = 0$ must be included and θ must range over say $-\pi < \theta \leq \pi$. However, the polar co-ordinates will then exhibit an edge at $\theta = \pi$ which does not exist according to Cartesian co-ordinates, or for that matter according to our usual ideas of edges of sets in the plane. The simplest way to avoid this inconsistency is to omit $r = 0$ and $\theta = \pi$, leaving a local polar co-ordinate system.

An atlas of local co-ordinate systems for a set is a family of local co-ordinate systems whose domains cover the whole set, and where any two of them when restricted to the intersection of their domains are differentiably related local co-ordinate systems. Whereas the plane admits global co-ordinate systems such as Cartesian systems, for surfaces like a sphere only local co-ordinate systems properly reflect their geometry. In order to co-ordinatise such sets in a way that allows consistent notions of tangent vectors and the linear operations upon them we require an atlas of co-ordinate systems. We can go further and consider all possible local co-ordinate systems which fit together as an atlas, and this is called a differentiable structure. A differentiable structure is just a maximal (or unenlargeable) atlas of local co-ordinate systems.

A manifold is a set endowed with a differentiable structure. However, none of the issues relating to atlases of co-ordinate systems seem to impinge on the the way in which manifolds enter into statistics. In the traditional study of manifolds most of the interest lies in their large-scale structure. A sphere and a torus clearly differ in their geometry in the large. Even where standard treatments of differential geometry explore the behaviour of such surfaces near a point, such as their curvature, it is often to establish a relationship with large-scale structure. However, the issues raised by statistics are those to do with determining a true parameter by, for example, some asymptotic means, all of which relates to distributions near to the true one. The focus therefore remains on the small-scale behaviour, and this can be dealt with within a single co-ordinate system. What is important therefore, is not the concept of a manifold in the large, as reflected by atlases of co-ordinate systems, but simply the concepts of calculus on manifolds.

For this reason we have not introduced manifolds in the usual way, which is to begin with them as large-scale objects, specifically as topological spaces, and impose atlases of co-ordinate systems upon them which are consistent with their topology. In our approach a manifold acquires a topology in any case. It is the one in which a subset is declared open if it is a union of subsets with open co-ordinate images. Two standard requirements must must then be imposed. We require that our manifolds be Hausdorff, which means that any two points must lie in disjoint open sets, or equivalently that one can find at least one differentiable function which takes different values upon them. We also require that our manifolds be paracompact, which means that any covering by open sets, such as a covering by domains of local co-ordinate systems, can be refined so that each point lies in an open set that intersects only a finite number of members of the covering. This has an impact on patching arguments which are necessary for the construction of certain functions, differential forms and vector fields on the whole manifold. We reiterate that such issues, which arise from large scale considerations, appear to us not to be of such importance to statistics. We refer the interested reader to Koboyashi and Nomizu (1963).

Example 2.2.13 *Euclidean space.* Ordinary Euclidean space $\mathbf{R}^n$ is a manifold with the co-ordinates

$$x^i(p^1, \ldots, p^n) = p^i$$

Example 2.2.14 *Vector spaces.* If V is a finite-dimensional vector space then choosing a basis $v^1, \ldots, v^r$ defines co-ordinates θ by

$$v = \sum_i \theta^i(v) v^i$$

Changing the basis just changes the co-ordinates by a matrix transformation so the new co-ordinates are clearly smoothly compatible with the co-ordinates θ. This makes V a manifold.

Example 2.2.15 *Affine spaces.* If X is an affine space then any two affine co-ordinate systems are smoothly related so they make X into a manifold.

2.2.10 *Co-ordinates and geometry*

We conclude this chapter by illustrating some simple relationships between special co-ordinates and geometry for the three simplest manifolds, Euclidean space, vector space and affine space.

Consider a manifold P with an atlas consisting of one set of co-ordinates whose domain is all of P and whose image is all of $\mathbf{R}^n$. Then it is clear that we can identify it with $\mathbf{R}^n$. This is the simplest kind of manifold. Now complicate things a little by considering an atlas whose co-ordinates again have domain all of P and image all of $\mathbf{R}^r$, but require that for any two co-ordinates θ and χ there is a *constant* matrix X^i_j such that for all i

$$\theta^i = \sum_j X^i_j \chi^j \tag{2.2.10}$$

Such a manifold arises if we consider a vector space V and the co-ordinates θ obtained by choosing a basis $v_1, \ldots, v_n$ for V and defining

$$v = \sum \theta^i(v) v_i$$

for any v in V. We claim that the converse is also true. Namely any P with such an atlas is a vector space. Certainly we can identify P with $\mathbf{R}^n$, which is a vector space, by choosing some co-ordinates from the atlas. It then inherits a vector space structure from $\mathbf{R}^n$. Moreover if we change the co-ordinates the linear relation (2.2.10) implies that the vector space structure on P is unchanged.

Finally consider a manifold with an atlas all of whose co-ordinate are related by affine transformations. That is for any two set of co-ordinates θ and χ we have a matrix X^i_j and a vector ξ^i such that

$$\theta^i = \sum_j X^i_j \chi^j + \xi^i$$

We saw in section 1.3 that an affine space gave rise to such an atlas. We want to now prove the converse.

In order to see how to construct the vector space of translations for P let us examine how the translations of an affine space affect the functions which arise as components of affine co-ordinate systems. To begin with, given arbitrary functions θ on a set P and a transformation $T : P \to P$, consider the difference function $\theta(T(p)) - \theta(p)$. This difference, which we will denote $\tilde{T}(\theta)$, defines a linear transformation $\tilde{T}$ from the vector space of all real-valued functions on P to itself. Note that constant functions necessarily map to the zero function.

Let us now specialise this observation to an affine space P and consider only the subset of functions θ which arise as component functions for affine co-ordinate systems. The affine relatedness of these co-ordinate systems means that the component functions for any one such system lie in the linear span formed from the components of any other and the constant functions. So the component functions of affine co-ordinate systems along with the constant functions form a vector subspace W of the space of all functions. Any transformation $T : P \to P$ will give a linear map from W into the general space of functions. However, it is easy to see that if T is a translation and θ is in W then $\tilde{T}(\theta)$ is a constant function, i.e. the difference $\theta(T(p)) - \theta(p)$ is independent of the point p if T is a translation and θ arises from an affine co-ordinate system. We can therefore think of $\tilde{T}$ as a real-valued linear function on W which, as we have noted, has the value zero on constant functions. It is easy to show that every such real-valued linear function arises from a unique translation. In summary, the map $T \to \tilde{T}$ defines a linear bijection between the vector space of translations V of an affine space and the subspace of W^* consisting of real-valued linear functions on W which take the value zero on constant functions.

We can reverse these observations. Given a set P and a family of affinely related co-ordinate systems we first consider the vector space W spanned by the constant functions and all of the functions which arise as components of these co-ordinate systems. Then we consider all linear functions on this vector space which take the value zero on the constant functions. This latter vector space parametrises a family of transformations of P to itself, which are easily shown to satisfy the two conditions for an affine space. Indeed, if t is any such function then we interpret its value on one of the co-ordinate functions θ as the change $\theta(T(p)) - \theta(p)$ for every p. That is, we insist upon the identity

$$\theta(T(p)) = \theta(p) + t(\theta)$$

By letting θ vary over the components $\theta^1, \ldots \theta^r$ of a co-ordinate system this determines the co-ordinates of $T(p)$ and hence $T(p)$ itself.

2.3 Remarks for Chapter 2

Remark 2.1 For further reading there are lots of books on differential geometry. We list a few here that reflect our own upbringing rather that an exhaustive library search. For an account developed around the classical differential geometry of surfaces in $\mathbf{R}^3$ see Klingenberg (1978). For a leisurely account with lots of pictures see Spivak (1970). The traditional reference for modern differential geometry is Koboyashi and Nomizu (1963) which statisticians will probably find a little terse. For those interested in infinite-dimensional manifolds Lang (1967) is a short book which also does the theory of manifolds modelled on a Banach space. Helgason (1978) gives the usual material on manifolds and the theory of Lie groups which are relevant to transformation models.

2.4 Exercises for Chapter 2

Exercise 2.1 Let $f: P \to V$ be a function from a manifold P into a vector space V. If X is a tangent vector to p at P then we define

$$d_p(f)(X) = (f \circ \gamma)'(0)$$

where γ is any path with $\gamma'(0) = X$. We call $d_p f$ a '1-form with values in V'.

Consider the log-likelihood

$$\ell: P \to R_\Omega$$

for a parametrised family P. If θ are co-ordinates on P expand $d_p\ell$ in terms of the $d_p\theta^i$.

Exercise 2.2 Let P be a parametrised family and $f: P \to R_\Omega$ be a function. Then we can define a function

$$E(f): P \to \mathbf{R}$$

by $p \mapsto E_p(f)$. Ignoring questions of differentiability and assuming that we can differentiate through the integral prove that

$$d_p(E(f)) = E_p(f(p)d_p\ell) + E_p(d_p f)$$

CHAPTER 3

Statistical manifolds

3.1 Realising manifolds

Any parametrised family of probability distributions,

$$P = \{p(x, \theta)\}$$

with parameter θ running over some open subset of $\mathbf{R}^r$, is automatically a manifold, in which the probability distributions are the points of the manifold and the parameters are co-ordinate functions. However, this is not to say that the family can be regarded as some definite *surface*, or *hypersurface*, on which the parameters play the role of co-ordinates. A manifold can be realised as a surface or hypersurface in an infinite variety of ways. Consider, for example the α-embeddings of Amari (1985):

$$F_\alpha(p) = \begin{cases} \frac{2}{1-\alpha} p^{\frac{1-\alpha}{2}}, & \alpha \neq 1 \\ \log(p), & \alpha = 1 \end{cases}$$

which map P into R_Ω, the space of random variables.

In one sense therefore, when we speak about a particular manifold we might imagine ourselves to be speaking about a whole family of surfaces which we regard as equivalent. We do much the same thing in talking about a sphere, in that there are an infinity of such surfaces, each specified by its centre and radius, but regarded as equivalent when talking about a sphere in the abstract. In regarding the actual spheres as equivalent we are effectively making transformations from one to another, setting up correspondences between their points. Alternatively, in regarding them as realisations of some ideal sphere we are making a transformation from the points of the ideal sphere to its realisations. The kind of transformations which we allow in the case of the geometrical concept of a sphere are quite restricted, namely rotations and dilations. If we were interested in the sphere only as a manifold then a lot more transformations would be allowed, and a lot more surfaces would qualify as realisations of it, such as ellipsoids and other distortions of it.

The idea of a transformation from one manifold to another is central to many ideas about manifolds, although we have raised it here only in the context realising a manifold as a hypersurface, or as an equivalence class of such hypersurfaces. The precise concept is that of a smooth map from one manifold to another, and in the following section we shall spell out exactly what this means.

Classically many manifolds where first described as hypersurfaces in $\mathbf{R}^3$ such as the sphere. This leads more generally to the concept of a submanifold of another manifold. Ideally we would like to define a statistical manifold to be a submanifold of $\mathcal{P}$. However $\mathcal{P}$ is generally an infinite-dimensional manifold and the treatment of such things is beyond the scope of this book. Instead we define a statistical manifold essentially as the image of what is known in differential geometry as an injective immersion. For this reason we devote some time in section 3.1.4 to a discussion of injective immersions and their relationship with submanifolds. The definition of as statistical manifold then follows easily.

We also use smooth maps to introduce the concept of a Lie group and Lie group actions. These are the basis of the transformation models discussed by Barndorff-Nielsen, which include location scale models, the von Mises–Fisher model and its hyperbolic analogue, along with a number of others. Transformation models are homogeneous spaces of certain Lie groups and as such are direct generalisations of affine spaces. In this sense transformation models are the next geometric step after exponential families. Certain transformation models, such as the von Mises–Fisher model are also curved exponential models, and we are able to give a simple presentation of some results from Barndorff-Nielsen (1988) characterising them.

3.1.1 *Smooth maps*

A transformation from one set M to another set N is just a function which accepts any point of M and returns a point of N. Functions between sets are also called maps. The set M is called the domain of F and N is called the codomain. The notation $F: M \to N$ is used to denote a map from M to N. As an example we might consider M to be three-dimensional space, N to be a plane inside it, and the map, transformation, or function F to be orthogonal projection of points onto the plane N. Thus for any point p the value $F(p)$ is the point in the plane N which is closest to it.

Loosely speaking a map sets up a correspondence between points of its domain and codomain, but the correspondence is not necessarily perfect since many points in the domain may be made to correspond to any one point of the codomain, and some points in the codomain need not have any points corresponding to them at all. A map which sets up a perfect correspondence, in which each point q in the codomain corresponds to exactly one point in the domain, is called a *bijection*. If it sets up a less perfect correspondence, in which each point q in the codomain arises as the value at at least one (although maybe more) point, then it is called a *surjection*. If the map sets up a correspondence in which each point q of the codomain corresponds to no more than one point in the domain (although perhaps none) then it is called an *injection*. Notice that orthogonal projection onto a plane is a surjection but not an injection or a bijection. In fact, a map is a bijection if and only if it is simultaneously an injection and a surjection.

A map $F: M \to N$ also sets up a correspondence between real-valued functions on N and real-valued functions on M by composition. Any real-valued function $\phi: N \to \mathbf{R}$ produces a real-valued function $\phi \circ F$ on M. We often call $\phi \circ F$ the *pullback* of ϕ under (composition with) F, since the correspondence goes in the opposite direction from F itself, i.e. whereas F transforms points in M into points in N, composition with it transforms functions on N into functions on M.

If M and N are manifolds then a map $F: M \to N$ is called *smooth* if each smooth function ϕ on N pulls back under F to a smooth function on M. In short $\phi \circ F$ should be smooth whenever ϕ is. In testing for smoothness one usually applies the criterion first of all to the functions $\phi^1, \ldots, \phi^n$ of a co-ordinate system on N. If F is to be smooth then each of $\phi^1 \circ F, \ldots, \phi^n \circ F$ must necessarily be smooth functions on M. This necessary condition is also sufficient however, because any smooth function on N is by definition of the form $\bar{f} \circ (\phi^1, \ldots, \phi^n)$ where $\bar{f}$ is smooth, and the smoothness of its pullback under F follows from the identity

$$(\bar{f} \circ (\phi^1, \ldots, \phi^n)) \circ F = \bar{f} \circ (\phi^1 \circ F, \ldots, \phi^n \circ F)$$

Hence, in order to show the smoothness of a map F we need only show that the pullbacks of some co-ordinate system are smooth functions. The smoothness of maps is actually a direct generalisation of the concept of smoothness of variations, where a variation γ is called smooth, relative to a co-ordinate system $\phi^1, \ldots, \phi^n$, if the functions $\phi^1 \circ \gamma, \ldots, \phi^n \circ \gamma$ are smooth, which in turn implies the smoothness of $f \circ \gamma$ for every other smooth function f.

In describing maps and testing their smoothness, the same liberties and more are often taken as in describing variations and testing their smoothness. For example, a map from the plane into three-dimensional space is often described in the form

$$(x(u,v), y(u,v), z(u,v))$$

where (u, v) are Cartesian co-ordinates in the plane, and (x, y, z) are Cartesian co-ordinates in space. This is the exact analogue of describing a variation in space in the form $(x(t), y(t), z(t))$. Just as $x(t)$ means the x co-ordinate of some point $\gamma(t)$, and should be written $x \circ \gamma(t)$, so $x(u, v)$ means the x co-ordinate of some point $F(p)$ in space, where (u, v) are the co-ordinates of p. The map F has been completely suppressed from this notation, just as γ is suppressed from the description of the variation. However, we cannot make the map explicit just by replacing x by $x \circ F$, and likewise with y and z, because $x \circ F$ is a function on the plane while the x in $x(u, v)$ is a function of pairs of real numbers. What is intended is that $x \circ F(p)$ is given by $x\,(u(p), v(p))$, so that x in this second expression is actually being used to denote the co-ordinate form of $x \circ F$, i.e. $\overline{x \circ F}$.

Replacing x with $\overline{x \circ F}$ in the expression $(x(u,v), y(u,v), z(u,v))$, and similarly for y and z, re-instates the underlying spaces and the map F between them, as well as summarising the way in which the co-ordinate expression for the map arises. However, if the map is understood and the process of co-ordinatisation can be taken for granted then it is far more convenient to keep x, y and z as abbreviations for $\overline{x \circ F}$, $\overline{y \circ F}$ and $\overline{z \circ F}$. In particular, whenever a map is expressed in this way, it is precisely x, y, and z, now considered as functions of pairs of real numbers, whose smoothness is required for the smoothness of the map F. Although we have confined this story to maps from the plane into space, simply to fix our ideas, it can be told for maps between any pair of manifolds. If $\theta^1, \ldots, \theta^m$ is a co-ordinate system on a manifold M and $\phi^1, \ldots, \phi^n$ is one on N then a map $F: M \rightarrow N$ is often described by an expression $\phi^1(\theta^1, \ldots, \theta^m), \ldots, \phi^n(\theta^1, \ldots, \theta^m)$ in which ϕ^i is used ambiguously as a replacement for $\phi^i \circ F$. The smoothness of the map F is established by the smoothness of $\phi^1, \ldots, \phi^n$ as functions of m real numbers $\theta^1, \ldots, \theta^m$.

3.1.2 *The derivative map*

The most significant aspect of the theory of manifolds is the notion of tangent map. A smooth map $F: M \rightarrow N$ not only pulls functions back from N to M, it also pushes tangent vectors forward from M to N, just as it does points. Specifically, F induces a map $TF: TM \rightarrow TN$ whose restriction to T_pM for any point p is a linear map into $T_{F(p)}N$. The map TF induced on tangent vectors is called the *derivative map* of F.

The derivative map arises because variations are pushed forward just like points. For a map $F: M \rightarrow N$, a variation γ through a point p gives rise to a variation $F \circ \gamma$ through $F(p)$ by composition with F. If F is smooth then smooth variations push forward to smooth variations. Indeed, we can test the smoothness of $F \circ \gamma$ by considering the smoothness of $f \circ (F \circ \gamma)$ for each smooth function f on N. But this is the same as $(f \circ F) \circ \gamma$ and the function $f \circ F$ is smooth by the smoothness of F. So the smoothness of γ implies the smoothness of $(f \circ F) \circ \gamma$ for every smooth f, and hence of $f \circ (F \circ \gamma)$, and this says that $F \circ \gamma$ is smooth.

Velocities are understood as collections of variations γ through some point p, identified in that each smooth function has a rate of change at p which is the same along each variation in the collection. This applies in particular to those smooth functions $f \circ F$ which are pulled back under F from smooth functions on N. Since $f \circ (F \circ \gamma) = (f \circ F) \circ \gamma$ this shows that each smooth function f on N has a rate of change at $F(p)$ which is the same along each of the variations $F \circ \gamma$. In short, smooth variations with the same tangent vectors in M push forward to smooth variations with the same tangent vector in N. This is the map T_pF from tangent vectors in M to those in N. Notice also that

$$(f_1 + f_2) \circ F = f_1 \circ F + f_2 \circ F \quad \text{and} \quad (\lambda f) \circ F = \lambda(f \circ F)$$

for any smooth functions f, f_1 and f_2 on N and any real number λ. Following through the effects of these for rates of change along smooth variations shows that

$$T_pF(v_1 + v_2) = T_pF(v_1) + T_pF(v_2) \quad \text{and} \quad T_pF(\lambda v) = \lambda T_pF(v)$$

for any tangent vectors v, v_1, v_2 in T_pM and any real number λ. So the restriction of T_pF to any tangent space T_pM is a linear map, just like df.

As with any linear map, the derivative map $T_pF: T_pM \to T_qN$, where $q = F(p)$, is determined completely by its values on any basis of T_pM, and these values can be described by their expansions in terms of some basis of T_qN. Such bases are provided by the partial derivatives with respect to co-ordinate systems. If $\theta^1, \ldots, \theta^m$ is a co-ordinate system on a manifold M and $\phi^1, \ldots, \phi^n$ is one on N, then $d\phi^i\left(T_pF(\partial/\partial\theta^j)\right)$ is the matrix of coefficients of the expansion of $T_pF(\partial/\partial\theta^j), j = 1 \ldots m$ with respect to $\partial/\partial\phi^i, i = 1 \ldots n$. It is called the Jacobian matrix of F with respect to these co-ordinate systems. Note that $\partial/\partial\theta^j$ is the tangent vector of a variation γ under which θ^j has unit rate of change, while the other θ^i's have zero rate of change, and $T_pF(\partial/\partial\theta^j)$ is the tangent vector of $F \circ \gamma$. It follows that $d\phi^i\left(T_pF(\frac{\partial}{\partial\theta^j})\right)$ is the rate of change of $\phi^i \circ (F \circ \gamma)$. It is the same as the partial derivative of $\phi^i \circ F$ with respect to θ^j. As described above, F is often suppressed from these notations in favour of writing $\phi^1, \ldots, \phi^n$ as functions of $\theta^1, \ldots, \theta^m$, which give the ϕ co-ordinates of $F(p)$ in terms of the θ co-ordinates of p. In this case the Jacobian matrix is expressed as the partial derivative of ϕ^i with respect to θ^j.

3.1.3 *Submanifolds*

Submanifolds of $\mathbf{R}^n$ are a generalisation of the idea of smooth surfaces in space, including 'hypersurfaces' of various dimensions rather than just dimension $n-1$. To begin with, consider the subsets of $\mathbf{R}^n$ of the form

$$\{(x^1, \ldots, x^r, 0, \ldots, 0) \mid (x^1, \ldots, x^r) \in \mathbf{R}^r\} \tag{3.1.1}$$

which can be identified with $\mathbf{R}^r$ in the obvious way. This situation is the archetypal example of a submanifold. As is often the case in differential geometry we create the general concept by 'localising' this example. We say that a subset Q of a manifold P is a submanifold if around every point q in Q we can find co-ordinates θ on P such that the points of Q near q are precisely those for which $\theta^{r+1}, \ldots, \theta^n$ vanish. The first r co-ordinate functions $\theta^1, \ldots, \theta^r$ when restricted to Q define co-ordinates on Q which make it a manifold. Notice that these co-ordinates map P to $\mathbf{R}^n$ in such a way that Q maps to the $\mathbf{R}^r$ described in (3.1.1). We say that 'locally' Q inside P looks like $\mathbf{R}^r$ inside $\mathbf{R}^n$.

Example 3.1.1 *Curved exponential families.* If P is an exponential family then a submanifold Q of P is called a *curved exponential family.* For a curved exponential family Q we have a function

$$\theta: Q \to \mathbf{R}^n$$

such that

$$q = \exp(\sum_i x^i \theta^i(q) - K(q))$$

The difference between this and an exponential family is, of course, that θ is not a bijection.

A typical way of describing a submanifold of $\mathbf{R}^n$ is as the solution set of a system of equations. A system of equations on $\mathbf{R}^n$ is a list

$$\begin{gathered} f_1(x) = c_1 \\ \vdots \\ f_{n-r}(x) = c_{n-r} \end{gathered}$$

where $f_1, \ldots, f_{n-r}$ are smooth functions on $\mathbf{R}^n$ and $c_1, \ldots, c_{n-r}$ are real numbers. The set of solutions of such a system need not be a submanifold in general. For example the solutions of the equation $x^2 - y^2 = 0$ form two intersecting lines as in Figure 3.1. The point of intersection of these lines doesn't admit any co-ordinate system making it look like a piece of $\mathbf{R}$ inside $\mathbf{R}^2$.

In order to show that the solution set of a system of equations is a submanifold we must, in principal, find for each solution x a complementary list of functions $\phi_1, \ldots, \phi_r$ such that

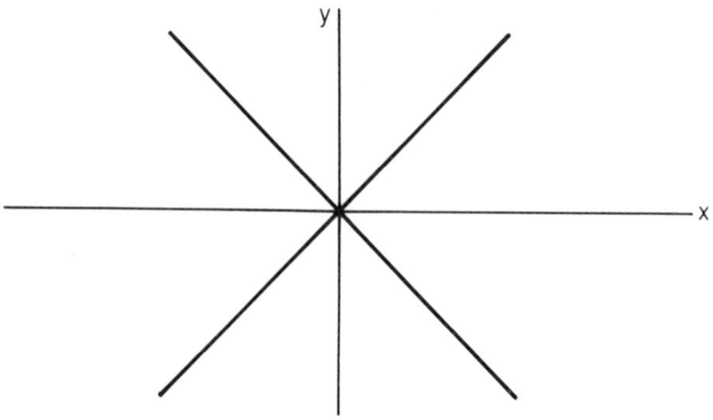

Figure 3.1

$\phi_1, \ldots, \phi_r, f_1 - c_1, \ldots, f_{n-r} - c_{n-r}$ are a co-ordinate system on a neighbourhood of x in $\mathbf{R}^n$. It is usually necessary to use a number of different complementary lists which operate in different regions of the solution set, i.e. it is not usually the case that one can find a complementary list which produces a co-ordinate system on the whole of $\mathbf{R}^n$ or even over the whole solution set. However, if such lists can be found then the solution set for the system of equations clearly satisfies the conditions required for it to be a submanifold of $\mathbf{R}^n$.

If $\phi = (\phi_1, \ldots, \phi_n)$ is a co-ordinate system on on some open subset U of $\mathbf{R}^n$ then it is a bijection $U \to V$ where V is an open set, and its inverse is also smooth. As a consequence $T\Phi$, the derivative map, is also a bijection on each tangent space. It is a fundamental result of multidimensional calculus, the inverse mapping theorem, that any smooth map $\Phi: U \to \mathbf{R}^n$ whose derivative map $T\Phi$ is a bijection on the tangent space T_xU at a point x has a smooth inverse when restricted to some neighbourhood of x, and hence produces a co-ordinate system on this neighbourhood. From this one can deduce that whenever a smooth map $F: U \to \mathbf{R}^{n-r}$ has a surjective derivative map $T_xF: T_xU \to T_y\mathbf{R}^{n-r}$, where $y = F(x)$, then a list of functions can be found to complement $F = (f_1, \ldots, f_{n-r})$ and make it into a co-ordinate system in a neighbourhood of x. This is called the submersion theorem.

Example 3.1.2 The classical method of applying the submersion theorem is to compute the Jacobian matrix of the map F and test that it has rank $n - r$ at each point x satisfying the equations

$F(x) = c$. For example, the Jacobian matrix of the map

$$F\begin{pmatrix} x \\ y \\ z \end{pmatrix} = \begin{pmatrix} x^2 + y^2 + z^2 \\ x + y + z \end{pmatrix}$$

is

$$\begin{pmatrix} 2x & 2y & 2z \\ 1 & 1 & 1 \end{pmatrix}$$

It has rank 2 unless $x = y = z$. Since there are no such points satisfying

$$\begin{aligned} x^2 + y^2 + z^2 &= 9 \\ x + y + z &= 3 \end{aligned}$$

the solutions of this system of equations forms a submanifold of $\mathbf{R}^3$.

Example 3.1.3 $SO(n)$. A less classical example, of some importance to us, is the space $SO(n)$ of orthogonal matrices of positive determinant. The set M of $n \times n$ matrices is identified with $\mathbf{R}^{n^2}$ via their lists of components. The set S of symmetric such matrices is identified with $\mathbf{R}^{n(n-1)/2}$ via their components above and on the diagonal. We regard the determinant as a real-valued function of matrices, which is smooth since it is a polynomial in the entries of a matrix. Consequently, the set of matrices on which det takes a positive value is an open set U in M. We define the map $F: U \to S$ by $F(A) = AA^t$, and $SO(n)$ is the set of solutions in U of the equation $F(A) = I$. To show that it is a submanifold, i.e. it has no singular points, we apply the submersion theorem.

Because M is a vector space for each tangent vector at a matrix A there is a matrix B such that the tangent vector is tangent to the variation $A + tB$. This is the canonical identification of $T_A M$ with M itself described in Chapter 2. Similarly the tangent space in S of any symmetric matrix A is identified with S. We compute the derivative map of F by considering

$$F(A + tB) = AA^t + tAB^t + tBA^t + t^2 BB^t$$

Identifying tangent vectors with pairs of matrices (A, B), as described above, it follows that

$$T_A F(A, B) = (AA^t, AB^t + BA^t)$$

For any matrix A satisfying $AA^t = I$, and symmetric matrix C, the equation

$$T_A F(A, B) = (I, AB^t + BA^t) = (I, C)$$

is solved by $B = 1/2CA$. Thus the derivative map is surjective at every A in $SO(n)$, and the submersion theorem implies that it is a submanifold of M.

3.1.4 *Inclusions, immersions and embeddings*

Curves and surfaces are often described in parametric form rather than as the set of solutions of a system of equations. More generally, one can attempt to describe r-dimensional submanifolds of a manifold P in terms of smooth maps $i: U \to P$ where U is an open subset of $\mathbf{R}^r$. The values $i(\theta)$ trace out a set Q of points in P as θ ranges over U. However, this set need not be a submanifold. To begin with we should require that different values of θ correspond to different points in P, i.e. that i is injective. In this case we can consider the parameters $\theta = (\theta^1, \ldots, \theta^r)$ to be functions of the points in $Q = i(U)$. However, in order that Q be a submanifold we would need these functions to extend to smooth functions on some open set containing Q, and to be complemented by other smooth functions $\theta^{r+1}, \ldots, \theta^n$ which are constant on Q and such that $\theta^1, \ldots, \theta^n$ is a co-ordinate system on P. If such complementary functions can be found in a neighbourhood of a point $q = i(\theta)$ then it is easy to show that the derivative map $T_\theta i$ is also injective on $T_\theta U$. Conversely, as a theorem of inverse function type, one can show that if the derivative map $T_\theta i$ is injective at a point θ then i itself is injective when restricted to some neighbourhood of θ and the image of this neighbourhood is a submanifold of P.

For a manifold Q a smooth map $i: Q \to P$ is called an immersion if its derivative map is injective at every point of Q. If i is an immersion then each point of Q has a neighbourhood whose image is a submanifold. However the image of Q as a whole need not be a submanifold for two basic reasons. First, i need not be injective. In this instance we think of Q as being identified with a subset of P via i, but in a way that causes Q to intersect itself. For example, we might immerse a circle in the plane as a figure 8, as in Figure 3.2.

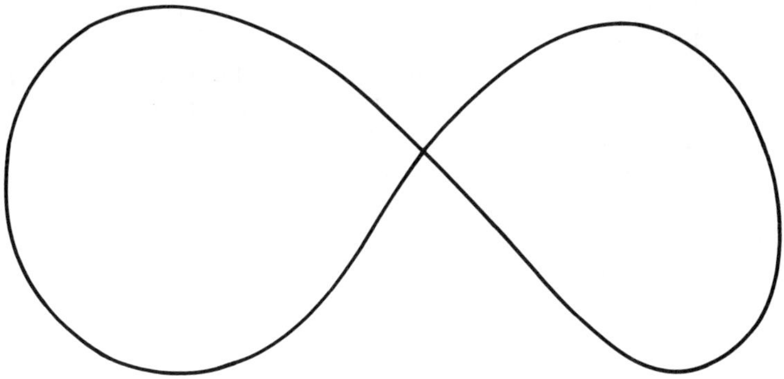

Figure 3.2

In order to avoid this problem we must require that the immersion i be injective. However, even an injective immersion need not have its image a submanifold. As an example, we might immerse the real line $\mathbf{R}$ into the plane as in Figure 3.3. This type of problem is avoided by placing a topological restriction on the immersion i, namely we require that it be a *proper* map. An injective proper immersion is called an *embedding*, and its image is guaranteed to be a submanifold. An embedding $i: Q \to P$ should be regarded as realising the manifold Q as a submanifold of P. It is not necessary for the purposes of this book to know what a proper map is, but the reader should appreciate the point that images of manifolds under quite reasonable maps may not be submanifolds.

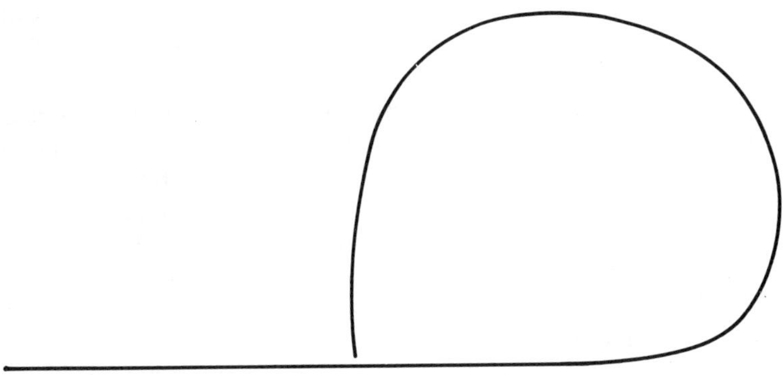

Figure 3.3

When a manifold Q arises as a submanifold of a manifold P it is useful to distinguish two different ways of thinking about it. Firstly we can think of the submanifold Q disembodied from P as a manifold in its own right. This is similar to thinking of a sphere in the abstract. Secondly the points of Q also happen to be points in P, and thinking of Q inside P is like thinking of a particular sphere in space. In effect Q the manifold realises itself as Q the 'hypersurface' in P. This point of view is encapsulated in the *inclusion* map

$$i: Q \to P$$

which maps a point p in the disembodied Q to the same point in P, so that $i(p) = p$. Any submanifold Q comes equipped with this natural embedding map into its containing manifold P. Composition of a function f on P with the inclusion i is just its restriction to Q. Since co-ordinate systems on Q are obtained as part of co-ordinate systems on P it follows that the restriction to Q of a smooth function on P will be a smooth function on Q. In particular, this says that the inclusion map i is indeed a smooth map.

The derivative map Ti of an embedding maps tangent vectors in Q to tangent vectors in P. Hence, not only does an embedding realise the points of Q as points of P, but it also realises tangent vectors in Q as tangent vector s in P. If Q is actually a submanifold of P, and defined by equations, then we can characterise the subspace of the tangent space of P which is identified with the tangent space of Q in terms of the derivative map of the equations. Specifically, suppose that Q is the set of solutions of the equation $F(p) = c$ where $F: P \to Z$ is some smooth map and c is a point of Z, for example, it may be that $Z = \mathbf{R}^n$. Then, for any p in Q, T_pQ is the subspace of T_pP consisting of tangent vectors v which satisfy $T_pF(v) = 0$. Indeed, if γ is any variation through p which lies in Q then $F(\gamma(t)) = c$ for every t, i.e. $F \circ \gamma$ is a constant variation and has tangent vector zero. Hence $T_pF(v) = 0$ for any tangent vector v in Q. Since T_pF is surjective in the case of a submersion this subspace of tangent vectors has dimension $\dim P - \dim Z$ which is also the dimension of Q. Hence T_pQ must coincide with the space of solutions of $T_pF(v) = 0$. In the case of $SO(n)$, since $F(A, B) = AB^t + BA^t$ the tangent space to $SO(n)$ at any matrix A is the space of matrices B for which this expression is zero. These are pairs (A, B) for which $AA^t = I$ and BA^t is antisymmetric.

If P is an affine space modelled on a vector space V then we can identify every tangent space to P with the vector space V, so the embedding of a submanifold Q has a tangent map

$$T_p i: T_p(Q) \to T_p(P) = V$$

which identifies T_pQ with a vector subspace of V. For example, if P is three-dimensional space then V can be taken as the vector space of arrows, identified by parallel translation, and every tangent vector at a point p has a unique variation $p+tv$ by which we identify it with the arrow v based at p. If Q is a sphere and p lies in Q then a tangent vector at p in Q is a family of variations within Q which we regard as sharing the same tangent vector . None of them are of the form $p + tv$, for such a variation traces out a straight line which cannot lie in the sphere. However, the variations in Q are also variations in space P by inclusion, and the variations in P which share the same tangent vector form a larger family which does contain a uniform variation. Thus we obtain the usual vision of a tangent vector to a sphere as an arrow which is geometrically tangent to it.

To spell out this correspondence in more detail, fix an origin p for P and define

$$\ell: P \to V$$

by $q = p + \ell(q)$. If γ is a variation in Q then it is also a variation in P and shares the same tangent vector as the uniform variation

$$t \mapsto q + t\gamma'(0)$$

where $\gamma'(0)$ is defined in Example 2.2.6. Applying ℓ we see that

$$\gamma'(0) = (\ell \circ \gamma)'(0)$$

In particular the co-ordinate vector fields

$$\frac{\partial}{\partial \theta^i}$$

map to

$$\frac{\partial \ell}{\partial \theta^i}$$

under the derivative map of the inclusion.

3.2 The definition of a statistical manifold

If $\mathcal{P}$ were a manifold we might just say that a statistical manifold is any finite-dimensional submanifold of $\mathcal{P}$. However $\mathcal{P}$ is usually an infinite-dimensional space and to make it an infinite-dimensional manifold would require a number of further topological considerations that would lead us too far afield from the geometry we wish to consider. So we will take the risk of defining a statistical manifold as basically the image of an injective immersion into $\mathcal{P}$, finessing the fact that $\mathcal{P}$ is not a manifold by using its affine structure. Our motivation is to regard a statistical manifold as a surface in $\mathcal{P}$.

Note that our definition of a statistical manifold is quite different to that of Lauritzen in Amari et al. (1987).

3.2.1 *Statistical manifolds and the score*

Let P be a subset of $\mathcal{P}$. We define a function

$$f: P \to R_\Omega$$

to be smooth if for every x the real-valued function

$$p \mapsto f(p)(x)$$

on P is smooth.

Then a statistical manifold is a subset of $\mathcal{P}$, the space of probability measures in a measure class $\mathcal{M}$ on a sample space Ω, which is also a manifold and satisfies two conditions:

1. Choosing any origin μ for $\mathcal{M}$, the log-likelihood function

$$\begin{array}{rcl} \ell: P & \to & R_\Omega \\ p = e^f \mu & \mapsto & f \end{array}$$

is a smooth function in the sense we have defined.

2. For any point p and co-ordinates θ about p the random variables

$$\frac{\partial \ell}{\partial \theta^i}(p)$$

are linearly independent.

Notice that (1) is true for any choice of origin if it is true for one particular origin, and that (2) is true for any co-ordinate system if it is true for any one co-ordinate system. The first of these statements is obvious and the second can be proved as follows. If we have other co-ordinates χ then we have

$$\frac{\partial \ell}{\partial \chi^j} = \frac{\partial \theta^i}{\partial \chi^j}\frac{\partial \ell}{\partial \theta^i}$$

and because the co-ordinates are smoothly compatible the Jacobian matrix

$$\frac{\partial \theta^i}{\partial \chi^j}$$

is invertible.

The first condition is essentially the requirement that the inclusion

$$P \subset \mathcal{P}$$

is smooth and the second is the requirement that it is an immersion. As we have seen in the discussion of submanifolds of affine spaces, its derivative map can be regarded as a linear map

$$d_p(\ell) \colon T_pP \to R_\Omega$$

and, in co-ordinates, it has the form

$$d_p(\ell)(\frac{\partial}{\partial \theta^i}) = \frac{\partial \ell}{\partial \theta^i}$$

We call this map the *score*. In the statistical literature the vector

$$(\frac{\partial \ell}{\partial \theta^1}, \frac{\partial \ell}{\partial \theta^2}, \ldots, \frac{\partial \ell}{\partial \theta^r})$$

is usually called the score vector, or just the score.

We shall often think of the score as a differential 1-form with values in R_Ω. This differential 1-form is

$$d_p\ell = \sum_i \frac{\partial \ell}{\partial \theta^i}(p) d_p\theta^i$$

In summary, a statistical manifold P is a family of probability measures for which the log-likelihood is a differentiable map and the score is an inclusion.

Example 3.2.1 *The score of the exponential family.* Consider an exponential family P. Then

$$\ell(p)(x) = \sum_i \theta^i x^i - K(p)$$

So the score is

$$d_p\ell = \sum_i x^i d_p\theta^i - d_p K$$

and the image of the tangent vector

$$\frac{\partial}{\partial\theta^i}$$

under the score is the random variable

$$\frac{\partial\ell}{\partial\theta^i} = x^i - \frac{\partial K}{\partial\theta^i}$$

Example 3.2.2 *The score of the normal family.* For a normal family with unknown variance we have

$$\ell = \frac{-(x-\mu)^2}{2\sigma^2} - \log(\sqrt{2\pi}\sigma)$$

and

$$d\ell = (\frac{x-\mu}{\sigma^2})d\mu + (\frac{(x-\mu)^2}{\sigma^3} - \frac{1}{\sigma})d\sigma$$

Example 3.2.3 *The score of the unit normal family.* Consider p in the unit normal family $\mathcal{N}(\mathbf{R}^n, 1)$ so that

$$p(\mu)(x) = \frac{1}{(2\pi)^{n/2}}\exp(-\frac{1}{2}|x-\mu|^2)$$

Choose as origin

$$\mu = \frac{1}{(2\pi)^{n/2}}\exp(-\frac{|x|^2}{2})dx^1\ldots dx^n$$

then

$$\ell(p)(x) = \langle x,\mu\rangle - \frac{1}{2}\langle\mu,\mu\rangle$$

and

$$d\ell = \sum_i (x^i - \mu^i) d\mu^i$$

Example 3.2.4 *The score of a curved exponential family.* Let Q be a curved exponential family with densities

$$q = \exp(\sum_i x^i \theta^i(q) - K(q))$$

Then if χ are local co-ordinates on Q we have

$$d_q \ell = \sum_j (\sum_i x^i \frac{\partial \theta^i}{\partial \chi^j} - \frac{\partial K}{\partial \chi^j}) d_q \chi^j$$

3.2.2 *Some useful formulae*

We have already used the identity

$$E_p(\frac{\partial \ell}{\partial \theta^i}) = 0 \tag{3.2.1}$$

obtained by differentiating

$$1 = E_p(1) = \int_\Omega \exp(\ell(p))\mu \tag{3.2.2}$$

Let us collect here two other similar identities for later reference.

The first is obtained by differentiating (3.2.1) to give

$$E_p(\frac{\partial \ell}{\partial \theta^i} \frac{\partial \ell}{\partial \theta^j}) = -E_p(\frac{\partial^2 \ell}{\partial \theta^i \partial \theta^j}) \tag{3.2.3}$$

For the second let $f: P \to R_\Omega$ be a differentiable function. We can then define a function $E(f)$ on P by $p \mapsto E_p(f(p))$ and we have

$$\frac{\partial}{\partial \theta^i} E(f) = E(\frac{\partial \ell}{\partial \theta^i} f) + E(\frac{\partial f}{\partial \theta^i}) \tag{3.2.4}$$

Notice that we are assuming here that we can differentiate through the integral sign. This is another question relating to the infinite-dimensional differentiable structure of $\mathcal{P}$ and we again choose to ignore it.

3.2.3 *The second fundamental form of a family*

We now have the concepts necessary to give an invariant definition of the α_{ij} as promised at the end of Chapter 1. Recall that there we defined an inner product $\langle , \rangle_p$ on R_Ω by

$$\langle f, g\rangle_p = E_p(fg)$$

Assume for simplicity that $E_p(f^2) < \infty$ for all random variables in R_Ω. Let N_p be the orthogonal to $T_p\tilde{P}$ inside R_Ω with respect to the inner product $\langle , \rangle_p$. Denote by π_p the orthogonal projection map from R_Ω to T_pP which has kernel N_p. As we saw in Chapter 1 this can be described explicitly by

$$\pi_p(f) = f - \sum g^{mn} E_p(f\ell_m)\ell_n - E_p(f)$$

If X and Y are vector fields on P define

$$\alpha(p)(X, Y) = \pi_p(X(Y(\ell)))$$

Notice that $\alpha(p)(X, Y)$ depends on the vector field X only via its value at p but, at first sight, appears to depend on the vector field Y by its values in a neighbourhood of p as we need to differentiate $Y(\ell)$. In fact if we calculate in co-ordinates it is easy to see that $\alpha(p)(X, Y)$ depends only on the value of Y at p. We have

$$\begin{aligned}\alpha(p)(X, Y) &= \pi_p(X^iY^j\frac{\partial^2\ell}{\partial\theta^i\partial\theta^j} + X^i\frac{\partial Y^j}{\partial\theta^i}\frac{\partial\ell}{\partial\theta^j}) \\ &= \pi_p(X^iY^j\frac{\partial^2\ell}{\partial\theta^i\partial\theta^j})\end{aligned}$$

because π_p vanishes on the scores

$$\frac{\partial\ell}{\partial\theta^i}$$

So we conclude that we have a well-defined bilinear, symmetric, map

$$\alpha: T_pP \times T_pP \to N_p$$

for every p. It is pure linear algebra now to show that this is determined by its coefficients with respect to (a co-ordinate) basis. These are

$$\alpha(p)(\frac{\partial}{\partial\theta^i}, \frac{\partial}{\partial\theta^j}) = \pi_p(\frac{\partial^2\ell}{\partial\theta^i\partial\theta^j}) = \alpha_{ij}$$

The formulae in Chapter 1 can now be understood as describing the way that the coefficients of such a linear map will change if we change the basis.

Let us complete the discussion in Chapter 1 by showing that the condition for a statistical manifold to be exponential is that the second derivative of the log-likelihood, with respect to any parameters, should be a linear combination of the scores and the constant random variables.

Recall that in the discussion in section 1.5.1 we extended the statistical manifold P to a manifold $\tilde{P}$ of positive measures defined by

$$\tilde{P} = \{\exp(\lambda)p \mid \lambda \in \mathbf{R}, p \in P\}$$

We also extended the local co-ordinates θ^i by defining $\theta^i(\exp(\lambda)p) = \theta^i(p)$ and defined a new co-ordinate by $\theta^0(\exp(\lambda)p) = \lambda$. It is then easy to see that P is an affine subspace of $\mathcal{P}$ generated by a vector space $V \subset R_\Omega/\mathbf{R}.1$ if and only if $\tilde{P}$ is an affine subspace of $\mathcal{M}$ generated by

$$\tilde{V} = \{f \in R_\Omega \mid f + \mathbf{R}.1 \in V\}$$

So we noted that P is an exponential family if and only if $\tilde{P}$ is an affine subspace of $\mathcal{M}$.

We then observed the following condition that all affine subspaces must have. If μ is chosen as an origin for $\tilde{P}$ then every other measure in $\tilde{P}$ has the form $\exp(\tilde{\ell})\mu$ where $\tilde{\ell}$ ranges over the finite-dimensional subspace $\tilde{V}$ of random variables. Let $\tilde{\ell}(\exp(\lambda)p) = \ell(p) + \lambda$ be the log-likelihood generalised to this situation and let $\tilde{\ell}_i$ denote the ith partial derivative of $\tilde{\ell}$. Notice that by our definitions, $\tilde{\ell}_0$ is the constant random variable 1. Since all of the values of $\tilde{\ell}$ lie in the subspace $\tilde{V}$ so do the values of $\tilde{\ell}_0, \ldots \tilde{\ell}_r$ and moreover, as we have seen, they form a basis for $\tilde{V}$.

If P is affine then the partial derivatives of the score vectors must also lie in the space of translations $\tilde{V}$. If we denote

$$\frac{\partial^2 \tilde{\ell}}{\partial \theta^i \partial \theta^j}$$

by $\tilde{\ell}_{ij}$ then, since $\tilde{\ell}_0, \ldots \tilde{\ell}_r$ is a basis for $\tilde{V}$, there must be coefficients $\gamma_{ij}^k(\theta)$ such that

$$\tilde{\ell}_{ij}(\theta) = \sum_{k=0}^{r} \gamma_{ij}^k(\theta)\tilde{\ell}_k(\theta)$$

Also as $\tilde{\ell}_{0i}(\theta) = 0$ and $\tilde{\ell}_0(\theta) = 1$ it follows that an equivalent way of stating this condition is that the second derivatives of the usual log-likelihood ℓ_{ij} are in the span of the usual scores ℓ_i and the constant random variables.

We want to prove the claim of section 1.5.1, namely that this condition that the derivatives of the scores lie in the span of the scores at each point, is characteristic of affine subspaces. Indeed, let f be a linear function on R_Ω which is zero on the subspace spanned by the score vectors at one point, say $\theta = 0$, which we take to be the base measure from which densities and log-likelihoods are calculated. In order to save notation (by the usual abuses) define

$$f_i(\theta) = f(\tilde{\ell}_i(\theta))$$

for $i = 0, \ldots, r$. Then we have

$$\begin{aligned}\frac{\partial}{\partial \theta^j}(f_i(\theta)) &= \left(f\tilde{\ell}_{ij}(\theta)\right)\\ &= f\left(\sum_{k=0}^{r} \gamma_{ij}^k(\theta)\tilde{\ell}_k(\theta)\right)\\ &= \sum_{k=0}^{r} \gamma_{ij}^k(\theta) f_k(\theta)\end{aligned}$$

This means that the functions f_i, considered as functions of the single variable θ^j with the other θ's held fixed, satisfy a linear system of ordinary differential equations.

Let us assume that the coefficient functions γ_{ij}^k are sufficiently smooth (Lipschitz for example) that the existence and uniqueness theorem for solutions of such equations applies. Since $f_i(0, \ldots, 0) = 0$ it follows, using $j = 0$, that $f_i(\theta^0, 0, \ldots, 0) = 0$ for every θ^0. This in turn implies, using $j = 1$ that $f_i(\theta^0, \theta^1, 0, \ldots, 0) = 0$ for all θ^0 and θ^1. Continuing in this fashion shows that $f_i(\theta) = 0$ for every θ. Hence any linear function which vanishes on $\tilde{\ell}_0(0), \ldots, \tilde{\ell}_r(0)$ also vanishes on $\tilde{\ell}_0(\theta), \ldots, \tilde{\ell}_r(\theta)$ for every θ. It follows that $\tilde{\ell}_0(\theta), \ldots,$ $\tilde{\ell}_r(\theta)$ lie in the subspace spanned by $\tilde{\ell}_0(0), \ldots, \tilde{\ell}_r(0)$ for every θ, and since we are assuming the scores to be linearly independent functions at each point we may simply say that $\tilde{\ell}_0(\theta), \ldots, \tilde{\ell}_r(\theta)$ span the same subspace $\tilde{V}$ for every θ. Furthermore, if f is a linear function vanishing on $\tilde{V}$ then $f(\tilde{\ell}(\theta))$ has all of its partial derivatives $f_i(\theta)$ equal to zero for every θ. Since $\tilde{\ell}(0) = 0$ it follows that $f(\tilde{\ell}(\theta)) = 0$ for every θ, and so the log likelihood $\tilde{\ell}(\theta)$ itself lies in the subspace $\tilde{V}$ spanned by the scores at any one point. Therefore $\tilde{P}$ is an affine subspace of its measure class, so that P is an affine subspace of the class of probability measures, and thus an exponential family.

3.3 Curved exponential families

We can generalise the result in section 1.5 to curved exponential families. Recall from example 3.1.1 that a parametrised family of distributions Q is a curved exponential family if it is contained in an exponential family P modelled on a finite-dimensional vector space V. For any co-ordinates χ on Q the random variables

$$\frac{\partial^2 l}{\partial \chi^i \partial \chi^j} \tag{3.3.1}$$

take values in $\tilde{V}$. This follows because the log-likelihood relative to an origin takes values in $\tilde{V}$. The second fundamental form at some point p in Q therefore takes its values in the normal space to $T_p\tilde{Q}$ inside $\tilde{V}$. In such a setting the second fundamental form is essentially a finite-dimensional object taking its values in this finite-dimensional normal space.

We can prove a generalisation of the result in section 1.5 which tells us when a family is curved exponential. First let us define $T_p^{(k)}$ to be the span inside R_Ω of the constant random variables and the partial derivatives of the log-likelihood up to order including k. So $T_p^{(1)}$ is the tangent space to $\tilde{P}$. In general the dimension of $T_p^{(k)}$ may change as we vary p but, of course it is always less than or equal to $p_k(d)$, the dimension of the space of all polynomial functions in d variables of degree less that or equal to k plus 1. We say that P is k regular if the dimension of $T_p^{(k)}$ is $p_k(d)$ for all p. This is equivalent to asking that all the derivatives of the log-likelihood up to order including k, and the constant random variables are linearly independent.

Assume that P is k regular. Let $N_p^{(k)}$ be the normal space in R_Ω to $T_p^{(k)}$ and define the kth second fundamental form

$$\alpha^{(k)}_{i_1,\ldots,i_{k+1}}(p) \tag{3.3.2}$$

to be the projection of

$$\ell_{i_1,\ldots,i_{k+1}} \tag{3.3.3}$$

onto $N_p^{(k)}$.

We claim that if P is k regular and $\alpha^{(k)} = 0$ then P is a curved exponential family. To prove this we apply the same technique as in section 1.5 to show that the derivatives of ℓ up to order k at any point q are in $T_p^{(k)}$ for some fixed p. Hence by regularity $T_p^{(k)} = T_q^{(k)} = \tilde{V}$ for all p and q and some $\tilde{V}$. The scores are therefore in $\tilde{V}$ at every point and it follows, as in section 1.5, that the log-likelihood takes its values in $\tilde{V}$ so that the family is a subfamily of the exponential family modelled on $\tilde{V}/\mathbf{R}.1$.

3.4 Lie groups

Given two sets M and N their Cartesian product is the set $M \times N$ of all pairs (p, q) where p belongs to M and q belongs to N. If both M and N are manifolds then $M \times N$ can be made into a manifold in a natural way. Indeed, if M is parametrised by co-ordinate functions $\theta^1, \ldots, \theta^m$ and N by co-ordinate functions $\phi^1, \ldots, \phi^n$ then we can endow a point (p, q) in $M \times N$ with parameters $\theta^1(p), \ldots, \theta^m(p), \phi^1(q), \ldots, \phi^n(q)$. Thus the list $\theta^1, \ldots, \theta^m, \phi^1, \ldots, \phi^n$ can be considered as functions on $M \times N$, where the value of θ^i on any pair (p, q) is its value on p and of ϕ^i is its value on q, and they serve as co-ordinates on $M \times N$. It is quite straightforward to check that smoothly related co-ordinate systems on M and N combine in this way to give smoothly related co-ordinate systems on $M \times N$. These co-ordinate systems make $M \times N$ into a manifold, called the *product manifold* of M and N.

While this method of producing co-ordinate systems and hence a manifold structure for $M \times N$ may seem only natural, it also has two key properties which mark it out as important mathematically. First the projection maps

$$\begin{aligned} M \times N &\longrightarrow M \qquad & (p, q) &\mapsto p \\ M \times N &\longrightarrow N \qquad & (p, q) &\mapsto q \end{aligned}$$

are smooth maps when this differentiable structure is used for $M \times N$. Secondly, any two maps $F: X \to M$ and $G: X \to N$

determine a map $F \times G: X \to M \times N$ by the rule $F \times G(x) = (F(x), G(x))$. If $M \times N$ is given its natural differentiable structure then $F \times G$ is a smooth map from the manifold X whenever F and G are smooth. We will leave the proofs of these facts to the interested reader.

There are a number of familiar mathematical concepts that can be described in terms of product manifolds and maps between them. For example, any (finite-dimensional) vector space V is a manifold as described in example 2.2.14. Addition of vectors is an operation that applies to pairs of vectors and produces from them a single vector. Therefore it makes sense to regard addition as applying to elements of $V \times V$ to give another element of V. It is a map $V \times V \to V$, and if $V \times V$ is the product manifold then addition is a smooth map. We could write it $+: V \times V \to V$, although one doesn't usually write the sum of v and w as $+(v, w)$ except in certain books on computer science. If X is an affine space with V as its vector space of translations then translation of a point p in X by a vector v is an operation concerning such pairs (v, p). Such a pair determines a point in X, namely the translate of p by v. Hence translation applies to elements of $V \times X$ to produce elements of X. It is a map $+: V \times X \to X$, and if X is a manifold via its affine co-ordinate systems and $V \times X$ is the product manifold then the translation map $+$ is smooth.

Any map $F: M \times N \to N$ allows us to think of the elements of the set M as acting as transformations from N to itself. A fixed element p in M corresponds to the transformation of N which sends q to $F(p, q)$. We say that M operates or acts on the set N (via F). Often the F and all other notational paraphernalia is omitted, and one writes simply pq for the result of the map applied to the pair (p, q). That lets us think of p simultaneously as an element of M and as an operation on N, as an object waiting to be applied to an element q of N to produce another element of N. A vector space, under addition, is an example of a set acting upon itself, although one anyway writes $v + w$ rather than vw. If X is the set of square matrices of some fixed dimension then one usually does write just AB for their product, which is a map $X \times X \to X$. Another example which will prove significant for location scale models in statistics is an action of $\mathbf{R} \times \mathbf{R}^+$ upon itself, $\mathbf{R}^+$ being the positive real numbers, which is given by

$$(\mu_1, \sigma_1)(\mu_2, \sigma_2) = (\mu_1 + \sigma_1\mu_2, \sigma_1\sigma_2) \qquad (3.4.1)$$

These three examples, a vector space under addition, multiplication of matrices and the location scale action of $\mathbf{R} \times \mathbf{R}^+$ upon itself, are important basic examples of a set acting upon itself. It should also be noted that in each case the set is actually a manifold, and that the action is a smooth map. In what follows we will restrict the space of matrices to those which are invertible. Being an open set, this subset is still a manifold with matrix entries still playing the role of co-ordinates.

When a set acts upon itself, via some map $M \times M \to M$, it is possible to make sense of the requirement that $p(qr) = (pq)r$. When this relationship holds the action is called an *associative multiplication* or associative product. Let us remind the reader that when we have such an associative multiplication and the action of each element p is a bijection, and hence invertible, and that inverse is realised by the action of an element of M, then M is called a *group.* The element which realises the inverse action of p is denoted p^{-1}, and it necessarily satisfies $pp^{-1} = p^{-1}p = e$.

Example 3.4.1 *Location scale group.* We leave the test of the associativity of (3.3.1) to the reader. A clever way to proceed, however, is to identify the element (μ, σ) with the 2×2 matrix

$$\begin{pmatrix} \sigma & \mu \\ 0 & 1/\sigma \end{pmatrix}$$

and show that the product of two elements in the location scale action corresponds to the product of their associated matrices. Then the associativity of the location scale product is a consequence ofthe associativity of the product of matrices. Moreover, the identity matrix corresponds to the element $(0, 1)$, which one can also show directly to act as the identity under the location scale product. In the same vein one can derive, either by considering inverse matrices or directly, that $(-\mu/\sigma, 1/\sigma)$ is the inverse element of (μ, σ). It follows that the location scale action makes $\mathbf{R} \times \mathbf{R}^+$ a group, called the location scale group.

Example 3.4.2 *Vector groups.* The addition operation on a vector space V is associative. Moreover, a vector space is required as part of its definition to have an element 0 which acts as an identity operation under addition, and corresponding to each element v an element $-v$ such that $v + (-v) = 0$. Thus a vector space under addition is a group, called a vector group. Vector groups are distinguished from other groups in that $v + w = w + v$, i.e. the product operation is commutative. For other groups it is usually not the case that $pq = qp$, as for example in the location scale group, or in the matrix groups described next. In fact, if a Lie group has a commutative product then the simple topological conditions of connectedness and simple connectedness are enough to force it to be a vector group, so commutativity is a very strong restriction.

Example 3.4.3 *Matrix groups.* Matrix multiplication is associative, and there is an identity matrix for this operation. By confining ourselves to the invertible matrices we obtain a group.

As we have noted, the action in each example above is a smooth map between manifolds. Furthermore, the map which sends each element to its inverse is also a smooth map between manifolds. Groups G which are manifolds, whose multiplication $G \times G \to G$ is a smooth map, and whose inverse $G \to G$ (i.e. $p \mapsto p^{-1}$) is a smooth map, are called *Lie groups*, after Sophus Lie whose researches introduced this concept. The examples above are our basic examples of Lie groups.

3.4.1 *Lie group actions*

As its name suggests, a Lie group action is an action $G \times X \to X$ of a Lie group G on a manifold X which is a smooth map. In addition, however, we require associativity, that is, $g(hx) = (gh)x$ for every g and h in G and x in X. It means that the operation of h upon X composed with the operation of g is realised by the operation of the element gh, so that once again composition of operations coincides with the product of elements of the group.

Example 3.4.4 *Affine space.* An affine space is an action of a vector group V upon a manifold X. It is distinguished from the most general example of a vector group action only in the requirement that any pair of points of X should have a unique translation which moves the first into the second.

Example 3.4.5 *Location scale action.* The location scale group $G = \mathbf{R} \times \mathbf{R}^+$ derives its name from its action on real numbers, in which an pair (μ, σ) scales a real number by σ and translates the result by μ. Hence the action $G \times \mathbf{R} \rightarrow \mathbf{R}$ is given by

$$(\mu, \sigma)x = \mu + \sigma x$$

Note that

$$(\mu_1, \sigma_1)\left((\mu_2, \sigma_2)x\right) = \mu_1 + \sigma_1(\mu_2 + \sigma_2 x) = (\mu_1 + \sigma_1\mu_2) + (\sigma_1\sigma_2)x$$

This shows that the action has the associativity required for it to be a Lie group action. It should also be said that the group multiplication for G was inspired by this requirement.

Example 3.4.6 *Location scale action on densities.* Given a function f of a single real variable x we can use an element of the location scale group as a change of variable to produce a new function, namely

$$(\mu, \sigma)f(x) = f\left(\frac{x-\mu}{\sigma}\right)$$

Note that

$$\begin{aligned}(\mu_1, \sigma_1)\left((\mu_2, \sigma_2)f\right)(x) &= (\mu_2, \sigma_2)f\left(\frac{x-\mu_1}{\sigma_1}\right) \\ &= f\left(\frac{\frac{x-\mu_1}{\sigma_1} - \mu_2}{\sigma_2}\right) \\ &= f\left(\frac{x-(\mu_1+\sigma_1\mu_1)}{\sigma_1\sigma_2}\right)\end{aligned}$$

It follows that the location scale group acts on the (infinite-dimensional) vector space of all functions. If we let X be the set of functions of precisely the form $f\left((x-\mu)/\sigma\right)$ for a chosen f, and let (μ, σ) be co-ordinates for this family, then the action of G on X is a Lie group action.

We can make this into an action on probability densities $f(x)dx$ simply by rescaling the transformed functions, i.e. we define

$$(\mu, \sigma)f(x)dx = \frac{1}{\sigma}f\left(\frac{x-\mu}{\sigma}\right)dx$$

Choosing $f(x)$ to be $1/\sqrt{2\pi}\exp(-\frac{x^2}{2})$ shows that the family of normal densities

$$N(\mu,\sigma)(x) = \frac{1}{\sqrt{2\pi}\sigma}\exp\{\frac{-(x-\mu)^2}{2\sigma^2}\}$$

admits an action by the location scale group, namely

$$\begin{aligned}(a,b)N(\mu,\sigma)(x) =& \frac{1}{a}\frac{1}{\sqrt{2\pi}\sigma}\exp\{\frac{-(\frac{(x-b)}{a}-\mu)^2}{2\sigma^2}\} \\ =& N(a\mu+b,\sigma a)(x)\end{aligned}$$

There is an analogy with affine spaces which is present in the action of the location scale group upon the normal family, which is not present in some of its other actions. Namely, given any pair of normal distributions there is a unique element of the location scale group which transforms the first into the second. This is not the case for the action of the location scale group on the vector space of all densities, nor for its action on the real line. In general, given any two univariate densities there will not be any element of the location scale group which transforms one into the other. In the case of the action on the real line one can always find an element of the location scale group which will move any one given point into another, however there are always many such elements. In particular, the uniqueness requirement analogous to that of affine spaces implies that no element other than the identity can move a point to itself, or rather leave it fixed. However, if we consider the action of the location scale group on the point 0, we see that (μ,σ) sends it to μ, so that all the elements of the group of the form $(0,\sigma)$ fix 0.

Example 3.4.7 *Representations.* Let us consider the action of a Lie group G on a vector space V. The most basic example is where G is the group of invertible $n \times n$ matrices acting on $\mathbf{R}^n$ by left multiplication. In this case, each transformation of V determined by an element of G, i.e. each map $v \mapsto gv$, is a linear transformation. An action of a group G on a vector space by linear transformations is called a *representation* of G.

Representations, that is, linear actions of groups, can be generated from any given action on a set X by using the functions on X. As we pointed out earlier, a transformation which moves points from a set X to a set Y pulls functions back from Y to X under composition. Given any element g of G we consider the pull back of functions on X under g^{-1}, in order to keep the direction the same. That is, for any function f on X we define gf to be the function

$$gf(x) = f(g^{-1}x)$$

Because each g acts on functions essentially by a change of variables it is easy to see that the sum of functions is transformed to the sum of their individual transforms, and likewise with scalar multiples. Hence one always produces a representation of G upon the real-valued functions on a set X upon which G acts by this device. The reader should check that this is exactly the process by which the action of the location scale group on functions was derived from its action on the real line.

Let us consider the action of the group of orthogonal 3×3 matrices on $\mathbf{R}^3$. By treating $\mathbf{R}^3$ as the Cartesian co-ordinates of points in space this is a model for the rotations of space about a fixed centre (the origin of the co-ordinate system). There are certain features of this action which are visualisable in this particular example, but which can be discerned in any group action. Any particular point of space is rotated about the origin by elements of the rotation group, and the set of all translates of a particular point x by all of the elements of the group form the sphere centred on the origin which contains x. In general, for a group G acting on a set X, a set of points $\{gx | g \in G\}$ obtained by operating on a particular element x with all of the elements of G is called an *orbit* of G. So the orbits of the rotation group acting on 3-space are the spheres centred on the origin of the rotation action. If we restrict ourselves to any one orbit X then we have a G action on X such that any pair of points has at least one element of g transforming the first into the second. When G acts on a set in this way the action is called *transitive*. One can always produce transitive G actions by restricting to the orbits of a given G action. The rotation group acts transitively on spheres.

We have already employed these ideas in relation to the action of the location scale group. Starting from its action on the real line, we produce its action on univariate functions and densities. These are representations, much like the action of the orthogonal matrices on $\mathbf{R}^3$. The normal densities form an orbit of this action.

Thus the location scale group acts transitively on the set of normal densities.

3.4.2 *Transformation models*

Suppose that a group G acts on a sample space Ω by measurable transformations. That is, Ω comes equipped with a σ-algebra of subsets, the events of the sample space, and any element g of the group maps events to events, i.e. the image of any subset in the σ-algebra under the transformation g also belongs to the σ-algebra. Just as we can make G act on the functions, or random variables, on Ω, so we can make it act on the probability measures. If μ is a measure on the σ-algebra, then $g\mu$ is the measure defined by

$$g\mu(E) = \mu\left(g^{-1}(E)\right)$$

for each event E. A *transformation model* is a family P of probability measures which is an orbit of such a G action. In particular, this is a transitive action of G on P. More generally, one could consider families of probability measures consisting of families of orbits, and not just a single orbit. These are called composite transformation models.

Example 3.4.8 *Location scale models.* If g is the element (μ, σ) of the location scale group then g^{-1} sends a real number x to $y = (x - \mu)/\sigma$. It follows that $dy = dx/\sigma$, so that g transforms any probability distribution $f(y)dy$ into $(1/\sigma)f\left((x - \mu)/\sigma\right)dx$. This family, parametrised by μ and σ, is the location scale model determined by the density f. It is the orbit of $f(y)dy$ under the action of the location scale group, and is therefore a transformation model for that group.

Example 3.4.9 *Von Mises–Fisher model.* Let S^{n-1} be the unit sphere in $\mathbf{R}^n$ and let G be the group $SO(n)$. The von Mises–Fisher model is the family of distributions on S^{n-1} parametrised by $\xi \in \mathbf{R}^n$, and given by

$$a(|\xi|)\exp(\xi.x)d\mu$$

where x is in S^{n-1}, μ is the area measure on S^{n-1}, and $a(|\xi|)$ is a normalisation factor determined by the requirement that the total mass of the distribution is one. If g belongs to $SO(n)$ then, since μ is rotationally invariant, we have

$$\mu(g^{-1}E) = \mu(E)$$

for every measurable subset of the unit sphere, and

$$\xi . g^{-1} x = \xi . g^{t} x = g\xi . x$$

The action of a rotation g on a distribution of the von Mises–Fisher family therefore gives

$$a(|\xi|) \exp(\xi . g^{-1} x) d(g\mu) = a(|\xi|) \exp(g\xi . x) d\mu$$

Thus the von Mises–Fisher model is a composite transformation model. Since $SO(n)$ acts on the parameter space $\mathbf{R}^n - \{0\}$ by rotations its orbits are the spheres, i.e. the subsets on which $|\xi|$ is constant. These subsets of the Von Mises–Fisher family are therefore transformation models. The value of $|\xi|$ is called the precision of the model.

The von Mises–Fisher model is clearly in the form of an exponential family, so it is simultaneously an exponential family (curved) and a (composite) transformation model. These are called *exponential transformation models* Barndorff-Nielsen (1988). A number of such models exist, namely the Dimroth–Watson and matrix von Mises–Fisher models, the Hyperboloid model analogous to the von Mises–Fisher model, and the Wishart and Bingham models Barndorff-Nielsen (1988). The univariate normal family is a location scale model which is also an exponential model.

In the Von Mises-Fisher model the densities of the family are taken with respect to a measure which is invariant under the action of the group, namely the surface area of the unit sphere. Such exponential transformation models relate precisely to finite-dimensional representations of the group. Indeed, given any distribution of the form $\exp(f)d\mu$ the image under the induced action of a group element g is $\exp(gf)d(g\mu)$, where $gf(x) = f(g^{-1}x)$ and $g\mu(E) = \mu(g^{-1}E)$. In an exponential family all of the functions gf must lie in a finite-dimensional subspace, and so the space which they themselves span must be a finite-dimensional subspace of random variables invariant under the induced action of G. Thus we have the following from Barndorff-Nielsen (1988).

Theorem *Let P be an exponential family, and therefore of the form* $\exp(f)d\mu$ *where f belongs to some finite-dimensional subspace V of random variables. Let P also be a transformation model under the action of some Lie group G on the sample space, and suppose that μ is an invariant measure, i.e. $g\mu = \mu$ for every g in G. Then V is a subspace of the space of random variables which is invariant under the induced action of G. Conversely, given any finite-dimensional subspace of random variables V which is invariant under the action of G, a family* $\exp(f)d\mu$ *consisting of finite measures in which f ranges over a G orbit in V is an exponential transformation model.*

For a great many Lie groups the finite-dimensional invariant subspaces of their induced actions on spaces of functions are well known and extensively studied. These finite-dimensional invariant subspaces are, in particular, representations of the group on a finite-dimensional subspace. For compact groups, semisimple groups, solvable groups and many others, these finite-dimensional representations have all been classified. Furthermore, there are theorems, such as those of Frobenius reciprocity type, which describe those finite-dimensional representations arising as subspaces of functions on a given manifold X upon which the group acts, which are invariant under its induced action. In the case of $SO(n)$ acting on the unit sphere, the restriction of harmonic polynomials to S^{n-1} which are homogeneous of some fixed degree constitute such invariant subspaces. They are minimal in the sense that they do not properly contain any other invariant subpaces. Furthermore, any finite-dimensional subspace of functions on the sphere which is invariant under $SO(n)$ must be spanned by these subspaces. The spaces of harmonic polynomials can therefore be used to construct further examples of exponential transformation models on the sphere, in addition to the von Mises–Fisher model.

Even for exponential transformation models that do not contain an invariant measure, it is still possible in many instances to describe their structure comprehensively in terms of representations of the Lie group involved. If μ is any measure in an exponential transformation model then the others are the measures $g\mu$ where g ranges over the elements of the group. As members of an exponential family they must be expressible in the form $\exp(f(g,x)d\mu$, where $f(g,x)$ is a random variable as a function of x, and these functions span a finite-dimensional subpace V of random variables as g ranges over G. Let us observe that

$$\begin{aligned} hg\mu &= \exp(f(hg,x))d\mu \\ &= h(gd\mu) \\ &= h\left(\exp(f(g,x))d\mu\right) \\ &= \exp(hf(g,x))d(h\mu) \\ &= \exp\left(hf(g,x)+f(h,x)\right)d\mu \end{aligned}$$

so that

$$f(hg,x) = hf(g,x) + f(h,x)$$

Suppressing the x, we may think of f as a function on the group G, i.e. a function of g, with values in the subspace V of random variables, upon which G acts by its induced action. The identity above can be abbreviated

$$f(hg) = hf(g) + f(h)$$

Such a function is known as a *1-cocycle* of G with values in the representation V. One way to produce such cocycles is to choose some element of the representation space, in our case a random variable b from V, and set

$$f(g) = gb - b$$

It is easy to check that the 1-cocycle identity above must hold. Cocycles of this special form are called *1-coboundaries.*

It is a fundamental theorem on Lie groups that semisimplicity is equivalent to the property that any 1-cocyle on a finite-dimensional representation must be a 1-coboundary. Semisimple Lie groups include the groups $SO(p,q)$ of invertible matrices of order $p+q$ and determinant 1 which preserve the indefinite inner product of signature (p,q). They also include their complex analogues $SU(p,q)$ and the symplectic groups. For a definition we refer the interested reader to any text on Lie groups, for instance Helgason (1978). For such groups we have the following from Barndorff-Nielsen (1988).

Theorem *Let P be an exponential family, and therefore of the form* $\exp(f)d\mu$ *where f belongs to some finite-dimensional subspace V of random variables. Let P also be a transformation model* *under the action of some Lie group G on the sample space. If G is semisimple then, given some measure μ in the family, there is a random variable b in V such that the family has the form* $\exp(gb - b)d\mu$.

The subspace V is once again an invariant subspace of the space of random variables under the induced action of G. In this instance, since μ need not be invariant, V may have to contain the constant functions. In any event, the theorem shows that for semisimple groups G, their exponential transformation models are determined by its finite-dimensional representation theory. Even if G is not semisimple, it may have exponential transformation models in the form of the Theorem above, because it still may happen that the 1-cocyle determined by the model is a 1-coboundary. This happens in the case of the univariate normal family, which is an exponential transformation model for the location scale group. The location scale group is solvable, not semisimple, and has 1-cocycles which are not 1-coboundaries.

3.5 Remarks for Chapter 3

Remark 3.1 Our definition of a statistical manifold is really just a summary of the properties we want it to have, and deficient in a number of respects. Most importantly we haven't addressed the question of how many of the derivatives of the log-likelihood we expect to be integrable. This is related to the fact that our definition of a smooth function into R_Ω is not very good. In particular there is not reason why the partial derivatives of a smooth function are also smooth. Thes questions could be properly addressed by considering the theory of Banach manifolds or perhaps Fréchet manifolds. The interested reader should consult Eells (1966) or Lang (1967).

3.6 Exercises for Chapter 3

Exercise 3.1 Let P be a statistical manifold and N an integer. Denote by Ω^N the product measure space. Recall that if p is a measure in P then there is a product measure p^N on Ω^N. Let P_N be the set of measures on Ω^N obtained by taking such products for all measures in P. If f is a random variable in R_Ω denote by f^N the random variable

$$f^N(x^1, \ldots, x^N) = f(x^1)\ldots f(x^N)$$

(1) Show that $(p+f)^N = p^N + f^N$.
(2) If ℓ_N is the log-likelihood for P_N show that $\ell_N(p^N) = \ell(p)^N$.
(iii) If θ are local co-ordinates on P define local co-ordinates on P_N by $\theta^i(p^N) = \theta^i(p)$. Show that

$$E_{p^N}(\frac{\partial \ell_N}{\partial \theta^i}\frac{\partial \ell_N}{\partial \theta^j}) = N E_p(\frac{\partial \ell}{\partial \theta^i}\frac{\partial \ell}{\partial \theta^j})$$

Exercise 3.2 Prove that

$$E_p(\frac{\partial \ell}{\partial \theta^i}\frac{\partial \ell}{\partial \theta^j}) = -E_p(\frac{\partial^2 \ell}{\partial \theta^i \partial \theta^j})$$

Exercise 3.3 Let $GL(n)$ denote the general linear group on $\mathbf{R}^n$, that is the group of all linear maps from $\mathbf{R}^n$ to $\mathbf{R}^n$ or n by n matrices. Show that an element X in $GL(n)$ acts on $\mathcal{N}(\mathbf{R}^n)$ and transforms the measure $p = p(\mu, \Sigma)$ into the measure $p(X\mu, X^{-1t}\Sigma X^{-1})$. See exercise 1.3 for the notation.

CHAPTER 4

Connections

4.1 Introduction: connections and geometry

Recall that a vector field is a choice at each point of a manifold of a vector in the tangent space at that point. The concept of the rate of change of a vector field has a profound relationship to geometry. This is because a straight line is a curve traced out by a point moving with constant velocity. The succession of velocities which a moving point adopts constitutes a vector field along the curve traced out during the course of its motion. We call this vector field the velocity field along the curve. The velocity field of a constant velocity motion is a constant vector field, and so these particular vector fields must be considered to have rate of change zero. If one considers the possibility of different geometries on a plane, so that the straight lines in these new geometries are not the same as the usual Euclidean straight lines, then our idea of a constant velocity motion must be changed to one that will trace out the new kind of straight line. Consequently the whole specification of the rate of change of vector fields must alter, since even the basic idea of what has rate of change zero must be tailored to produce the different families of straight lines.

Not only does the rate of change of vector fields alter if we change the family of straight lines, so do other vectorial quantities that relate to vector fields. In particular the rate of change of 1-forms such as the differentials of functions must be re-interpreted if certain natural relationships are to be maintained. Specifically, we have by definition that

$$dx(\frac{\partial}{\partial x}) = 1$$

and so

$$dx(\frac{\partial}{\partial x})$$

is a constant function, irrespective of any changes to the notion of straight line. While

$$\frac{\partial}{\partial x}$$

is considered a constant vector field we would want to regard dx as a constant 1-form in the light of this relationship. But if

$$\frac{\partial}{\partial x}$$

is no longer to be regarded as constant, because we are working in a new geometry, then we may prefer not to regard dx as constant, but as varying in some way that will maintain the constancy of

$$dx(\frac{\partial}{\partial x})$$

From the point of view of statistical theory, the possibility of altering the rate of change of 1-forms, by a change in geometry or whatever, is of great significance. The rate of change of the 1-form df pertains to the second derivatives of f, and altering it changes Taylor expansions of functions to second and higher orders. In this way geometry has has a direct influence on the form of asymptotic expansions.

The relationship between the rate of change of df, the second derivatives of f and Taylor expansion occurs through the formation of the bilinear function of tangent vectors

$$\nabla df(v, w) = \sum_{i,j=1}^{n} \frac{\partial^2 f}{\partial \theta^i \partial \theta^j}(p) d\theta^i(v) d\theta^j(w)$$

Given any variation γ through a point p this bilinear function allows the Taylor expansion of $f \circ \gamma$ about $t = 0$ to second-order to be expressed as

$$f(p) + df(v) + \frac{1}{2}\nabla df(v, v) + \ldots$$

where v is the tangent vector determined by γ. Fixing v, we interpret $\nabla df(v, w)$ as the value of the linear function $\nabla df(v, -)$

on w, and this linear function is itself interpreted as the rate of change of df under the velocity v. Thus $\partial^2 f/\partial\theta^i \partial\theta^j$ is interpreted as the rate of change of df along $\partial/\partial\theta^i$ evaluated on $\partial/\partial\theta^j$.

However, different co-ordinate systems $\theta^1, \ldots, \theta^n$ produce quite different bilinear forms ∇df, i.e. different real values $\nabla df(v, w)$ for the same choices of f, v and w. By contrast, for a given f and v the formula

$$\sum_{i=1}^{n} \frac{\partial f}{\partial\theta^i} d\theta^i(v)$$

produces the same value, namely $df(v)$, no matter what the choice of co-ordinate system θ. Different co-ordinate systems produce the same bilinear forms ∇df only if they are affinely related. Consequently the concept of the rate of change of df and second-order Taylor expansion, as described above, automatically determine an affine geometry.

As we shall see, there are more general geometries among which the affine geometries are regarded as the very simplest. In these more general geometries we can still define the rate of change of df and second-order Taylor expansion by the same co-ordinate formulas, but we have to change the co-ordinate system from point to point rather than use the one co-ordinate system overall. It is not a novel idea to use co-ordinate systems which are adapted to a particular point. Indeed, one often singles out a particular point in the plane to be the origin of a Cartesian co-ordinate system. More generally if a particular point p in a set is of special interest then one often requires of a co-ordinate system θ that $\theta(p) = 0$. The idea of adapting co-ordinate systems to each different point is actually quite natural in statistics. Often in statistics we have a preferred point, the true distribution p, for which we may want to choose co-ordinates with special properties. It may only be to require that $\theta(p) = 0$, but we could make more stringent requirements. Such an approach to second and higher derivatives leads to the notion of *co-ordinate strings*, which generalise the ideas that we have canvassed here.

4.2 Rates of change of vector fields on the plane

Before considering how to redefine the rate of change of a vector field on the plane, let us recall its customary definition. In classical vector analysis a vector field V on the plane is usually described as an expression $a(x,y)\mathbf{i}+b(x,y)\mathbf{j}$ where $\mathbf{i}$ and $\mathbf{j}$ denote the unit length vector fields in the directions of the x and y axes respectively. The partial derivatives of V are then defined to be

$$\frac{\partial V}{\partial x} = \frac{\partial a}{\partial x}\mathbf{i} + \frac{\partial b}{\partial x}\mathbf{j}$$

$$\frac{\partial V}{\partial y} = \frac{\partial a}{\partial y}\mathbf{i} + \frac{\partial b}{\partial y}\mathbf{j}$$

We shall use the notation

$$\frac{\partial}{\partial x} \quad \text{and} \frac{\partial}{\partial y}$$

instead of $\mathbf{i}$ and $\mathbf{j}$ and we consider the total differential of V defined by

$$\nabla V(v) = da(v)\frac{\partial}{\partial x} + db(v)\frac{\partial}{\partial y} \tag{4.2.1}$$

Making these changes may alter our perspective to that of manifolds, but it doesn't really change the substance of this prescription for the rate of change of a vector field on the plane. In particular, notice that

$$\frac{\partial}{\partial x} \quad \text{and} \quad \frac{\partial}{\partial y}$$

themselves are given rate of change zero by these formulae. This accords with our intuition about the Euclidean geometry of the plane, since a 'picture' of each of the vector fields

$$\frac{\partial}{\partial x} \quad \text{and} \quad \frac{\partial}{\partial y}$$

looks like a family of parallel arrows of unit length. Because of this basic example, constant vector fields are often referred to as parallel vector fields.

Taking (4.2.1) as the definition of the rate of change of a vector field, we should observe the following three basic properties.

1. $\nabla V(v)$ is linear as a function of tangent vectors $v \in T_pP$ and takes its values in T_pP
2. $\nabla(V_1 + V_2) = \nabla V_1 + \nabla V_2$ for any two vector fields V_1, V_2
3. $\nabla(fV)(v) = df(v)V(p) + f(p)\nabla V(v)$ for any differentiable function f and every $v \in T_pP$.

(4.2.2)

These three rules are the direct analogues of the rules satisfied by functions, and if we are thinking of inventing any other definition of the rate of change of a vector field then we shall insist that it obey these three rules. We usually call (3) the Leibniz rule.

Suppose we find an operation ∇ on vector fields that obeys the three rules (4.2.2). If we express a vector field V in the form $V = a(\partial/\partial x) + b(\partial/\partial y)$ then these rules imply that, whatever the formula for ∇, for each $v \in T_pP$

$$\begin{aligned}\nabla V(v) &= da(v)\frac{\partial}{\partial x} + a(p)\nabla(\frac{\partial}{\partial x})(v) + db(v)\frac{\partial}{\partial y} + b(p)\nabla(\frac{\partial}{\partial y})(v) \\ &= da(v)\frac{\partial}{\partial x} + db(v)\frac{\partial}{\partial y} + a(p)\nabla(\frac{\partial}{\partial x})(v) + b(p)\nabla(\frac{\partial}{\partial y})(v)\end{aligned} \tag{4.2.3}$$

Hence any operation ∇ which is acceptable as a rate of change of vector fields, in the sense that it obeys the rules (4.2.2), produces the formula (4.2.1) from classical vector analysis along with two more terms, one involving

$$\nabla(\frac{\partial}{\partial x})$$

and the other involving

$$\nabla(\frac{\partial}{\partial y})$$

If it is our prejudice that the rates of change of

$$\frac{\partial}{\partial x} \quad \text{and} \quad \frac{\partial}{\partial y}$$

should be zero then we will require that these two expressions should be zero, and (4.2.3) will reduce to the classical formula. In other words, the classical formula is the only acceptable definition of the rate of change of vector fields on the plane which accords with the Euclidean assumption that

$$\frac{\partial}{\partial x} \quad \text{and} \quad \frac{\partial}{\partial y}$$

should be constant.

Now let us observe that in everything we have explained so far we could have substituted polar for Cartesian co-ordinates. Any vector field V can be expressed in the form $V = a(\partial/\partial r) + b(\partial/\partial\theta)$. If we define

$$\nabla V(v) = da(v)\frac{\partial}{\partial r} + db(v)\frac{\partial}{\partial \theta} \tag{4.2.4}$$

for every $v \in T_pP$ it is easy to see that, just as for (4.2.1), all three rules (4.2.2) are obeyed, so that ∇ should be regarded as acceptable for describing the rate of change of vector fields. However,

$$\nabla(\frac{\partial}{\partial r})$$

and

$$\nabla(\frac{\partial}{\partial \theta})$$

are both zero according to this formula. A picture of

$$\frac{\partial}{\partial r}$$

looks like a field of arrows of unit length pointing radially outwards from the origin of the polar co-ordinate system, while the arrows for

$$\frac{\partial}{\partial \theta}$$

point anticlockwise along the circles centred on this origin and have length equal to the radius of the circle on which they stand. Neither of them should be considered parallel, or constant, according to our usual Euclidean notions.

Even more striking is the fact that, since

$$\nabla(\frac{\partial}{\partial r}) \quad \text{and} \quad \nabla(\frac{\partial}{\partial \theta})$$

are both constant vector fields according to (4.2.4), the lines radiating from from the origin of the polar system and the circles centred on it must be considered straight lines! Indeed, a point moving radially outwards from the origin at unit speed has

$$\frac{\partial}{\partial r}$$

as its velocity field, while a point moving at constant speed around a circle centred on the origin has a multiple of

$$\frac{\partial}{\partial \theta}$$

as its velocity field, and both of these are now being regarded as constant vector fields. However, other lines and circles in the Euclidean sense are not to be regarded as straight. This may seem quite absurd until we deform the plane in the following way. In three-dimensional space let us turn the plane into a cone by fixing one circle centred on the origin, say the one of unit radius, and pressing down on the origin so as to tilt the lines radiating from it into a conical configuration. Next blow the origin out into a genuine hole and continue the tilting until we are left with a cylinder. The radial lines in the plane have now become the rulings on the cylinder, while the circles have become its circular cross-sections. These are now genuinely straight lines on the cylinder. If there are any doubts about it, then cut the cylinder along one of the rulings and roll it out flat. Nothing on the cylinder is distorted by this process, but the rulings and circular cross-sections turn into Euclidean straight lines, see Figure 4.1. In the light of this revelation we should simply regard the radial lines and concentric circles of the polar co-ordinate system as a representation of certain straight lines on the cylinder via the correspondence which we have described.

We should take the view that defining the rate of change of vector fields by the formula (4.2.4) has changed the shape of the plane. It has altered its Euclidean geometry into a cylindrical one. The fact that this cylindrical geometry itself can be unrolled into a Euclidean one raises some further interesting questions and issues.

It may seem very strange that so many different definitions of the rate of change of vector fields are possible, for one is tempted

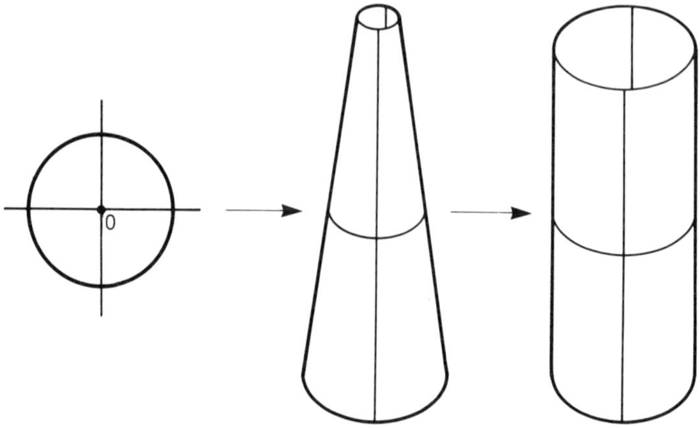

Figure 4.1

to imagine at first sight that 'rate of change' should be as unique a concept for vector fields as it is for functions. For a function f one begins with the difference between its values $f(p+tv) - f(p)$, divides by t and considers the limit as t approaches zero, so why not do the same for vector fields? The problem is that for a vector field V the values $V(p+tv)$ and $V(p)$ are tangent vectors located at two different points p and $p+tv$. In the plane we do not hesitate to subtract two such tangent vectors, we simply translate the one at $p+tv$, say, to p and use the parallelogram rule for subtraction there. But if we were on the surface of a cylinder how should we make the translation then? The most sensible idea for parallel translation of tangent vectors on the cylinder is the one obtained by rolling the cylinder out flat and then using the usual idea of parallel translation. If we allow that the plane can represent the geometry of the cylinder, via polar co-ordinates, then parallel translation of tangent vectors in the plane, done by the cylindrical prescription, will be utterly different from the normal Euclidean one. For example, in order to translate a tangent vector from one point on a circle centred on the origin to a different point on the same circle we would have to rotate it as we move it along the circle, keeping its angle with the radial direction the same all the time (Figure 4.2).

What emerges is that the seemingly insignificant step of translating a tangent vector from one point to another is profoundly

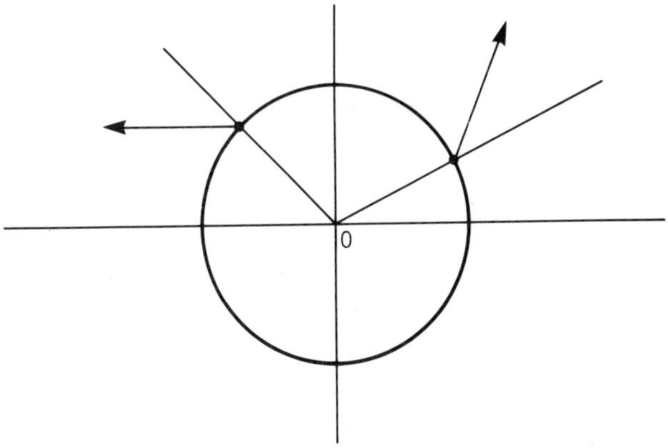

Figure 4.2

affected by the geometry of the plane which we are assuming. If we are to consider differences between tangent vectors such as $V(p+tv)$ and $V(p)$ then we must have a way of relating or connecting tangent vectors located at different points, and if there is no unique way to make such a connection then there can be no unique concept for the rate of change of a vector field.

The relationship of parallel translation and connection of tangent spaces to the concept of the rate of change of a vector field was a formative influence in differential geometry. Consequently mathematical gadgetry of any kind which allows one to specify the rate of change of vector fields is called a connection. The idea of specifying that any operation ∇ satisfying the conditions (4.2.3) should be interpreted directly as a rate of change of vector fields was first enunciated by J.L. Koszul in the 1950s. Any such operation is now called a *Koszul connection.*

Equation (4.2.3) shows how to evaluate a Koszul connection on any vector field once its values are specified on

$$\frac{\partial}{\partial x} \quad \text{and} \quad \frac{\partial}{\partial y}$$

This would be convenient if for some reason we were working in a Cartesian co-ordinate system, but if we were working in a polar one we would have the same formula as (4.2.3) with polar co-ordinates (r, θ) in place of (x, y), and we would need to specify the values of the connection on

$$\frac{\partial}{\partial r} \quad \text{and} \quad \frac{\partial}{\partial \theta}$$

Let us consider in general how to specify

$$\nabla(\frac{\partial}{\partial \phi^1}) \quad \text{and} \quad \nabla(\frac{\partial}{\partial \phi^2})$$

where (ϕ^1, ϕ^2) is a co-ordinate system on the plane. The value of each of them on a tangent vector $v \in T_pP$ must also lie in T_pP, and each tangent vector in T_pP can be expressed uniquely as a linear combination of

$$\frac{\partial}{\partial \phi^1}(p) \quad \text{and} \quad \frac{\partial}{\partial \phi^2}(p)$$

We can therefore express

$$\nabla(\frac{\partial}{\partial \phi^1})(v) \quad \text{and} \quad \nabla(\frac{\partial}{\partial \phi^2})(v)$$

in the form

$$\nabla(\frac{\partial}{\partial \phi^1})(v) = A_1^1(v)\frac{\partial}{\partial \phi^1}(p) + A_1^2(v)\frac{\partial}{\partial \phi^2}(p)$$
$$\nabla(\frac{\partial}{\partial \phi^2})(v) = A_2^1(v)\frac{\partial}{\partial \phi^1}(p) + A_2^2(v)\frac{\partial}{\partial \phi^2}(p)$$

We use the notation $A_i^j(v)$ to indicate the dependence of the coeficients on the tangent vector $v \in T_pP$. Thus A_i^j are real-valued functions on the tangent bundle, and it is easy to see that they are linear since

$$\nabla(\frac{\partial}{\partial \phi^1}) \quad \text{and} \quad \nabla(\frac{\partial}{\partial \phi^2})$$

are linear. Hence A_i^j are 1-forms, called the *connection 1-forms* of ∇ relative to

$$\frac{\partial}{\partial \phi^1} \quad \text{and} \quad \frac{\partial}{\partial \phi^2}$$

Since each 1-form can be expressed in the form $ad\phi^1 + bd\phi^2$ we can write

$$A_i^j = \Gamma_{i1}^j d\phi^1 + \Gamma_{i2}^j d\phi^2$$

The functions Γ^j_{ik} are called the *Christoffel symbols* of the Koszul connection ∇ with respect to the co-ordinate system (ϕ^1, ϕ^2). Notice that it follows from the definitions that

$$\nabla(\frac{\partial}{\partial \phi^k})(_{\frac{\partial}{\partial \phi^j}}) = \Gamma^i_{kj} \frac{\partial}{\partial \phi^i}$$

Example 4.2.1 *The cylindrical connection.* Let us calculate the connection 1-forms and Christoffel symbols in Cartesian co-ordinates for the cylindrical connection, i.e. the connection defined by the conditions

$$\nabla(\frac{\partial}{\partial r}) = 0 \quad \text{and} \quad \nabla(\frac{\partial}{\partial \theta}) = 0$$

We must calculate

$$\nabla(\frac{\partial}{\partial x}) \quad \text{and} \quad \nabla(\frac{\partial}{\partial y})$$

and express them as linear combinations of

$$\frac{\partial}{\partial x} \quad \text{and} \quad \frac{\partial}{\partial y}$$

Since we are given the value of ∇ on

$$\frac{\partial}{\partial r} \quad \text{and} \quad \frac{\partial}{\partial \theta}$$

namely zero, we should express

$$\frac{\partial}{\partial x} \quad \text{and} \quad \frac{\partial}{\partial y}$$

in terms of

$$\frac{\partial}{\partial r} \quad \text{and} \quad \frac{\partial}{\partial \theta}$$

We have

$$\begin{aligned} \frac{\partial}{\partial x} &= \cos\theta \frac{\partial}{\partial r} + \frac{-\sin\theta}{r} \frac{\partial}{\partial \theta} \\ \frac{\partial}{\partial y} &= \sin\theta \frac{\partial}{\partial r} + \frac{\cos\theta}{r} \frac{\partial}{\partial \theta} \end{aligned}$$

and so

$$\begin{aligned}\nabla(\frac{\partial}{\partial x}) &= d(\cos\theta)\frac{\partial}{\partial r} + d(\frac{-\sin\theta}{r})\frac{\partial}{\partial\theta}\\ \nabla(\frac{\partial}{\partial y}) &= d(\sin\theta)\frac{\partial}{\partial r} + d(\frac{\cos\theta}{r})\frac{\partial}{\partial\theta}\end{aligned} \tag{4.2.5}$$

From examples 2.2.5 and 2.2.6 we have

$$\begin{aligned}\frac{\partial}{\partial r} &= \cos\theta\frac{\partial}{\partial x} + \sin\theta\frac{\partial}{\partial y}\\ \frac{\partial}{\partial\theta} &= -r\sin\theta\frac{\partial}{\partial x} + r\cos\theta\frac{\partial}{\partial y}\end{aligned}$$

Substituting these expressions into (4.2.5) and simplifying gives

$$\begin{aligned}\nabla(\frac{\partial}{\partial x}) &= (\frac{-\sin^2\theta}{r})dr\frac{\partial}{\partial x} + (-d\theta + \frac{\sin\theta\cos\theta}{r}dr)\frac{\partial}{\partial y}\\ \nabla(\frac{\partial}{\partial y}) &= (d\theta + \frac{\sin\theta\cos\theta}{r}dr)\frac{\partial}{\partial x} + (\frac{-\cos^2\theta}{r})dr\frac{\partial}{\partial y}.\end{aligned} \tag{4.2.6}$$

The coefficients of

$$\frac{\partial}{\partial x} \quad \text{and} \quad \frac{\partial}{\partial y}$$

in (4.2.6) are by definition the connection forms of the cylindrical connection, expressed in terms of dr and $d\theta$. Expressing them in Cartesian terms gives

$$\nabla(\frac{\partial}{\partial x}) = (\frac{-y^2}{x^2+y^2})(\frac{xdx+ydy}{x^2+y^2})\frac{\partial}{\partial x} + (\frac{(y^3+2x^2y)dx - x^3dy}{(x^2+y^2)^2})\frac{\partial}{\partial y}$$

$$\nabla(\frac{\partial}{\partial y}) = (\frac{-y^3dx + (x^3+2xy^2)dy}{(x^2+y^2)^2})\frac{\partial}{\partial x} + (\frac{-x^2}{x^2+y^2})(\frac{xdx+ydy}{x^2+y^2})\frac{\partial}{\partial y}.$$

From here we can read off the Christoffel symbols.

4.3 Affine spaces and flat connections

Affine spaces have a natural notion of constant vector fields and hence a natural connection. If P is an affine space with V its vector space of translations, then at each point p in P the vector v determines a tangent vector, namely the velocity of the uniform variation $p + tv$. As explained in example 2.2.6, this association is a linear isomorphism of vector spaces $V \rightarrow T_pP$ for each point p. Thus each translation vector v determines a tangent vector at each point of P, which is to say that it determines a vector field, which we denote by the same letter. The natural connection arises by regarding these vector fields as constant. This is the direct generalisation of the situation in the plane, where a translation vector is an arrow, which produces an arrow at every point, namely its parallel translates, and these vector fields are regarded as constant. Regarding these vector fields as constant makes the constant velocity variations the same as uniform variations, so that straight lines are the curves traced out by uniform variations.

A choice of basis $v_1, \ldots, v_n$ for the space of translations provides a set of vector fields such that every other vector field X can be expressed

$$X = f_1v_1 + \ldots + f_nv_n$$

for some unique set of *coefficient functions* $f_1, \ldots, f_n$. We define

$$\nabla X(w) = df_1(w)v_1 + \ldots + df_n(w)v_n$$

∇ is a Kozsul connection, and the only one which is consistent with the requirement that the vector fields $v_1, \ldots, v_n$ be constant.

Another way to arrive at this definition is to notice that via the isomorphisms $V \rightarrow T_pP$ each vector field can be regarded as a function on P with values in V. Identifying vector fields with V valued functions in this way, we can define

$$\nabla X(w) = \frac{d}{dt}X(\gamma(t))\bigg|_{t=0}$$

where γ is any variation with velocity w. It is easy to see that this second definition gives the same result as the first, but it has the virtue of being independent of the choice of basis, and therefore shows that the connection ∇ is determined independently of any basis.

Let $\phi^1, \ldots, \phi_n$ be affine co-ordinates determined by a choice of origin and the basis $v_1, \ldots, v_n$. In other words, under the identification of P with V determined by the choice of origin, $\phi^1, \ldots, \phi_n$ are linear functions on V determined by the expansion

$$v = \phi^1(v)v_1 + \ldots + \phi_n(v)v_n$$

It follows at once that, regarded as vector fields, $v_1, \ldots, v_n$ are the co-ordinate vector fields $\partial/\partial\phi^1, \ldots, \partial/\partial\phi^n$. The natural connection ∇ is therefore characterised by the condition that $\nabla\partial/\partial\phi^i = 0, i = 1, \ldots, n$ for any affine co-ordinate system ϕ.

A Kozsul connection is called *flat* if there exists a co-ordinate system ϕ for which $\nabla\partial/\partial\phi^i = 0, i = 1, \ldots, n$. Such a co-ordinate system is called affine for the connection ∇. Any two such co-ordinate systems must be affinely related, and in this way a flat connection on a manifold determines a local affine structure, in the sense that each portion of the manifold covered by an affine co-ordinate system can be regarded as part of an affine space. For example, rolling a cylinder out onto a plane makes all of the cylinder minus the line along which it was cut, into part of an affine space, namely part of the plane.

Recall that the Christoffel symbols of a connection ∇ with respect to a co-ordinate system ϕ are the coefficient functions of expansion of $\nabla\partial/\partial\phi^i$, namely

$$\nabla\frac{\partial}{\partial\phi^i}\left(\frac{\partial}{\partial\phi^j}\right) = \sum_{k=1}^{n} \Gamma_{ij}^{k} \frac{\partial}{\partial\phi^k}$$

To say that the co-ordinate vector fields of some co-ordinate system are constant is obviously the same as to say that the Christoffel symbols with respect to this co-ordinate system are identically zero. Often differential geometers prefer to say that the Christoffel symbols of a connection 'vanish', i.e. they are identically zero, rather than be too informative and say simply that the co-ordinate vector fields are constant.

For two- or three-dimensional co-ordinate systems the ideas of treating them as affine co-ordinates, of defining a flat connection with them, and of determining an affine structure for which they are indeed affine co-ordinates, are commonplace. They are invoked implicitly by depicting the chosen co-ordinates as Cartesian co-ordinates. For example, the univariate normal family is often depicted via the (μ, σ) plane, so that the mean and standard deviation are treated as Cartesian co-ordinates. The **i** and **j** vector fields correspond to the co-ordinate vector fields $\partial/\partial\mu$ and $\partial/\partial\sigma$ and differentiating vector fields by simply differentiating the component functions of their expansions in terms of this basis, corresponds to using the natural flat connection. Translations in the plane produce corresponding translations of members of the normal family. All of this may seem quite obvious, and the fact that it is so obvious makes the notion of a flat connection in a way the most uninteresting kind of connection. Most people will say that while it is clear that depicting any co-ordinate system in a Cartesian way gives one the option of implanting a Cartesian geometry into their parametrised sets, they have never taken that geometric option seriously. Our point is that the minute they differentiate vector fields by the commonplace recipe, or make Taylor expansions to second order at varying points in their chosen co-ordinates, then they have taken that geometry very seriously without realising it.

4.4 Connections on submanifolds of affine spaces.

Consider the cylinder in the usual way, as a submanifold of three-dimensional space. A vector field on the cylinder is a vector field on a part of three-dimensional space, and if we can specify the rate of change of vector fields on three-dimensional space then why not use this to determine the rates of change of vector fields on the cylinder? More precisely, if we have a Koszul connection ∇ on three-dimensional space, such as the natural flat connection, then we can compute the rate of change $\nabla X(v)$ for any vector field X on three-dimensional space and for any velocity v. Why not apply this prescription to vector fields X restricted to the cylinder, using only those tangent vectors v which are the velocities of variations within the cylinder? If we could do this then any connection on three-dimensional space would produce a connection on the cylinder, or on any other surface.

The problem with this idea is that $\nabla X(v)$, if it is to be the rate of change of a vector field X on the cylinder, must itself be a tangent vector to the cylinder located at the same point as v. It is usually not. For example, if (r, θ, z) are cylindrical polar co-ordinates in three-dimensional space then cylinders are described by r = constant, and $\partial/\partial\theta$, $\partial/\partial z$ are vector fields along such cylinders. If ∇ is the natural flat connection on three-dimensional space, and (x, y, z) is a Cartesian co-ordinate system with the same z axis as the polar one, then we have

$$\nabla\frac{\partial}{\partial x} = 0 \qquad \nabla\frac{\partial}{\partial y} = 0 \qquad \nabla\frac{\partial}{\partial z} = 0$$

and the relations $x = r\cos\theta$, $y = r\sin\theta$ may be used to show that

$$\nabla\frac{\partial}{\partial r} = \frac{1}{r}d\theta\frac{\partial}{\partial\theta} \qquad \nabla\frac{\partial}{\partial\theta} = -rd\theta\frac{\partial}{\partial r} + \frac{dr}{r}\frac{\partial}{\partial\theta} \qquad \nabla\frac{\partial}{\partial z} = 0$$

Since r is constant on the surface of a cylinder we have $dr(v) = 0$ for any tangent vector v to the cylinder, and thus

$$\nabla\frac{\partial}{\partial\theta} = -rd\theta\frac{\partial}{\partial r}$$

where $d\theta$ is restricted to tangent vectors to the cylinder. While $\partial/\partial\theta$ is a vector field along the cylinder, $\partial/\partial r$ is not, and indeed points perpendicularly to it. Consequently the three-dimensional or ∇ rate of change of $\partial/\partial\theta$ has values perpendicular to the cylinder, not tangent to the cylinder as they should be in order to define a Koszul connection.

A simple modification of this procedure allows us to produce a connection on the cylinder. Any tangent vector can be expanded

$$w = dr(w)\frac{\partial}{\partial r} + d\theta(w)\frac{\partial}{\partial\theta} + dz(w)\frac{\partial}{\partial z}$$

Having computed $w = \nabla X(v)$, we can simply delete the $\partial/\partial r$ component, leaving a tangent vector to the cylinder. This is a kind of projection onto the tangent space of the cylinder, and indeed, since the polar co-ordinate vector fields are perpendicular at each point, considered as arrows in three-dimensional space, this operation corresponds precisely to our usual notion of othogonal projection onto the tangent spaces of the cylinder. Let us denote the operation of ∇ followed by projection by $\bar{\nabla}$. From the calculations above it follows that

$$\bar{\nabla}\frac{\partial}{\partial\theta} = 0 \qquad \bar{\nabla}\frac{\partial}{\partial z} = 0$$

In other words, this new connection $\bar{\nabla}$ is flat and has θ and z as affine co-ordinates. Replacing z by r, it is the same connection that we constructed rather arbitrarily by deciding that the polar co-ordinate vector fields in the plane should be constant. It is a remarkable fact about the geometry of the cylinder that, even though it is not flat, in the sense of being an affine subspace of three-dimensional space, the natural connection that it derives from it by projection is anyway flat, providing the cylinder with a local affine structure of its own.

Example 4.4.1 *Connection on the sphere.* Let us consider how to define a connection on the sphere, following the example of the cylinder. If (x, y, z) are Cartesian co-ordinates, and (r, θ, ϕ) are spherical polar co-ordinates with the same origin, then

$$x = r\cos\theta\cos\phi, \qquad y = r\sin\theta\cos\phi \quad \text{and} \quad z = r\sin\phi$$

From these relations we deduce

$$\begin{aligned}
\frac{\partial}{\partial r} &= \cos\phi\bigl(\cos\theta\frac{\partial}{\partial x} + \sin\theta\frac{\partial}{\partial y}\bigr) + \sin\phi\frac{\partial}{\partial z} \\
\frac{\partial}{\partial \theta} &= r\cos\phi\bigl(-\sin\theta\frac{\partial}{\partial x} + \cos\theta\frac{\partial}{\partial y}\bigr) \\
\frac{\partial}{\partial \phi} &= -r\sin\phi\bigl(\cos\theta\frac{\partial}{\partial x} + \sin\theta\frac{\partial}{\partial y}\bigr) + r\cos\phi\frac{\partial}{\partial z}
\end{aligned}$$

In order to follow the calculations more easily, let us introduce the vector field

$$V = \cos\theta\frac{\partial}{\partial x} + \sin\theta\frac{\partial}{\partial y}$$

Applying ∇, and also using the first and third of the equations above, we find

$$\begin{aligned}
\nabla V &= -\sin\theta d\theta\frac{\partial}{\partial x} + \cos\theta d\theta\frac{\partial}{\partial y} \\
&= \frac{d\theta}{r\cos\phi}\frac{\partial}{\partial \theta} \\
V &= \cos\phi\frac{\partial}{\partial r} - \frac{\sin\phi}{r}\frac{\partial}{\partial \phi} \\
\frac{\partial}{\partial z} &= \sin\phi\frac{\partial}{\partial r} + \frac{\cos\phi}{r}\frac{\partial}{\partial \phi}
\end{aligned}$$

Applying ∇ to the spherical co-ordinate vector fields gives

$$\begin{aligned}\nabla\frac{\partial}{\partial r} &= -\sin\phi d\phi V + \cos\phi\nabla V + \cos\phi d\phi\frac{\partial}{\partial z}\\ \nabla(\frac{1}{r\cos\phi}\frac{\partial}{\partial\theta}) &= -\cos\theta d\theta\frac{\partial}{\partial x} - \sin\theta d\theta\frac{\partial}{\partial y} = -d\theta V\\ \nabla\frac{1}{r}\frac{\partial}{\partial\phi} &= -\cos\phi d\phi V - \sin\phi\nabla V - \sin\phi d\phi\frac{\partial}{\partial z}\end{aligned}$$

Substituting the formulas for V and ∇V gives

$$\begin{aligned}\nabla\frac{\partial}{\partial r} &= \frac{d\phi}{r}\frac{\partial}{\partial\phi} + \frac{d\theta}{r}\frac{\partial}{\partial\theta}\\ \nabla(\frac{1}{r\cos\phi}\frac{\partial}{\partial\theta}) &= -\cos\phi d\theta\frac{\partial}{\partial r} + \frac{\sin\phi d\theta}{r}\frac{\partial}{\partial\phi}\\ \nabla(\frac{1}{r}\frac{\partial}{\partial\phi}) &= -d\phi\frac{\partial}{\partial r} - \frac{\tan\phi d\theta}{r}\frac{\partial}{\partial\theta}\end{aligned}$$

Now we must treat r as constant so as to restrict ourselves to the surface of a sphere, and therefore only consider tangent vectors v satisfying $dr(v) = 0$. Since the formulas involve no terms in dr this latter point has no effect on the calulations. In the formulas for $\bar{\nabla}$, we are only concerned with the terms in $\partial/\partial\theta$ and $\partial/\partial\phi$, so we delete the terms in $\partial/\partial r$. Hence we obtain

$$\begin{aligned}\bar{\nabla}\frac{\partial}{\partial\theta} &= -\tan\phi d\phi\frac{\partial}{\partial\theta} + \sin\phi\cos\phi d\theta\frac{\partial}{\partial\phi}\\ \bar{\nabla}\frac{\partial}{\partial\phi} &= -\tan\phi d\theta\frac{\partial}{\partial\theta}\end{aligned}$$

Notice that the spherical co-ordinates $\partial/\partial\theta$ and $\partial/\partial\phi$ are not affine co-ordinates for this connection. This raises the question as to whether the connection is flat or not, that is whether there is any co-ordinate system on the sphere whose co-ordinate vector fields are constant according to this connection. In Chapter 5 we shall introduce the notion of curvature in order to deal with this question, and show that there are no such co-ordinate systems.

Because the spherical co-ordinate vector fields are orthogonal, deleting the $\partial/\partial r$ terms is the same as orthogonal projection onto the tangent spaces of the sphere, just as in the case of the cylinder. Not all co-ordinate systems have orthogonal co-ordinate vector fields, and then deleting the unwanted co-ordinate terms will not correspond to orthogonal projection, but to some other type of projection onto the tangent spaces of the surface. In three-dimensional space orthogonal projection is considered the natural one, and the connection obtained by this method is called the *Levi-Civita* connection on the surface. One of its properties in particular is that the straight lines which it determines are also the curves which traverse the shortest distance between two points on the surface. Indeed, we shall show in Chapter 5 that the requirement that the shortest paths also be the straightest gives an intrinsic characterisation of the Levi-Civita connection.

In general, given a connection ∇ on a manifold P, any kind of projection process will produce a connection on a submanifold. More precisely, if Q is a submanifold then a projection onto its tangent bundle is a linear map $\pi_q : T_qP \to T_qQ$ at each point q in Q for which $\pi_q(v) = v$ whenever v lies in T_qQ. A projection allows us to define a connection on Q by the formula

$$\bar{\nabla} X(v) = \pi_q\big(\nabla X(v)\big)$$

To see that $\bar{\nabla}$ satisfies the rules (4.2.2) for a Koszul connection, notice that because π_q is linear the formula is linear in both X and v, so that only the Leibniz rule (number 3) remains to be verified. If X is a vector field on Q and f is a function then

$$\begin{aligned}\bar{\nabla}(fX)(v) &= \pi_q\big(df(v)X(n) + f(n)\nabla X(v)\big)\\ &= df(v)\pi_q\big(X(n)\big) + f(n)\pi_q\big(\nabla X(v)\big)\\ &= df(v)X(n) + f(n)\bar{\nabla} X(v)\end{aligned}$$

where the second line follows from the linearity of π_q and the third from the projection property. Thus $\bar{\nabla}$ is a Koszul connection as claimed.

Let us return to the natural flat connection ∇ on an affine space P and consider the case of an affine subspace Q. We should note that in this special case Q carries its own natural flat connection, and that any connection defined by projecting ∇ simply reproduces it. Indeed, if V is the space of translation of P, and W is the subspace corresponding to the space of translations of Q, then a vector field X on Q corresponds to a W-valued function. The derivative of this W-valued function, which produces $\nabla X(v)$, has values in W, and consequently, in this special case, $\nabla X(v)$ already lies in the tangent space of Q. Hence projections make no change to $\nabla X(v)$, so that all of the projections of ∇ simply reproduce the natural flat connection on Q.

4.4.1 *The second fundamental form*

We can extend the definition of the second fundamental form to this general setting. If we let Q_q be the kernel of the projection π_q then we have $T_qP = Q_q \oplus T_qQ$. The second fundamental form α of the connection with respect to the projection π_q is defined by

$$\alpha(X, Y) = (1 - \pi_q)\nabla Y(X)$$

As in the case of the second fundamental form of a family we have to check that α depends only on the values of the vector fields X and Y at the point q. It follows from the definition that

$$\nabla Y(X) = \tilde{\nabla} Y(X) + \alpha(X, Y)$$

The reader should beware a notational conflict between our notation here and that used in Chapter 1 for the second fundamental form of a family where π denoted the projection onto the *normal* space. We are now using π for the projection onto the tangent space so that the projection onto the normal space is $1 - \pi$.

To see what the second fundamental form looks like in co-ordinates let us choose co-ordinates $(\theta^1, \ldots, \theta^n)$ so that $(\theta^1, \ldots, \theta^r)$ define co-ordinates on Q and the remaining $\theta^{r+1}, \ldots, \theta^n$ all vanish on Q. That we can do this is the definition of a submanifold, see section 3.1. The projection operator π defines a matrix π^i_j by

$$\pi\left(\frac{\partial}{\partial \theta^j}\right) = \sum_{i=1}^{r} \pi^i_j \frac{\partial}{\partial \theta^i}$$

so that the second fundamental form is

$$\alpha_{ij} = \sum_{k=r+1}^{n} \Gamma^k_{ij}\left(\frac{\partial}{\partial \theta^k} - \sum_{m=1}^{r} \pi^m_k \frac{\partial}{\partial \theta^m}\right)$$

for $i, j = 1, \ldots, r$. Notice that the second fundamental form is symmetric if the connection is symmetric.

4.5 Amari's 1-connection

Let us return from general differential geometry to the theory of a statistical family. Amari (1985) defines a connection on any statistical family, called the 1-connection or exponential connection, which can be produced by the projection construction described above. Recall that a measure class $\mathcal{M}$ is an affine space whose vector space of translations is a space of random variables R_Ω. According to the general theory, any vector field on $\mathcal{M}$ can be represented as an R_Ω-valued function on $\mathcal{M}$. In particular, if P is a statistical manifold then a vector field on P can be represented as an R_Ω-valued function, whose values at each point are restricted to the subspace corresponding to the tangent space of P at that point. If P is parametrised then the components $\ell_1, \ldots, \ell_n$ of the score are examples of such vector-valued functions. They represent the co-ordinate vector fields $\partial/\partial\theta^1, \ldots, \partial/\partial\theta^n$, and they span the subspace of R_Ω corresponding to the tangent space of P.

Given a vector field X on $\mathcal{M}$, considered as an R_Ω-valued function, we define $\nabla X(v)$ as the t derivative of $X(\nu(t))$ at $t = 0$, where $\nu(t)$ is a 1-parameter variation with velocity v. That is, $X(\nu(t))$ is a family of random variables parametrised by t, hence a function of t and a sample point, and its t derivative means its partial derivative with respect to t. If P is a parametric family of probability distributions then we might choose X to be the vector field ℓ_i and v to be $\partial/\partial\theta^j$ at some particular point, so that $\nabla X(v)$ is just the t derivative of $\ell_i(x, \theta^1, \ldots, \theta^j + t, \ldots \theta^n)$, i.e the jth partial derivative of ℓ_i. In Chapter 1 we denoted these partial derivatives by ℓ_{ij}. These functions do not usually lie in the span of the scores. Indeed the main result of Chapter 1 was that the functions ℓ_{ij} lie in the span of the scores ℓ_i and the constants at every point if and only if the statistical manifold is an exponential family.

In general a parametric family of probability distributions is a submanifold of $\mathcal{M}$ in much the same way as a sphere is a submanifold of three-dimensional space. In both cases, in order to produce a connection on these submanifolds from the natural connection on their containing affine space, we need a way of projecting the values $\nabla X(v)$ onto their tangent spaces. In three-dimensional space we had available the usual orthogonal projection. In the space of measures $\mathcal{M}$ we have a substitute provided by the inner products on R_Ω

$$\langle f, g \rangle_p = E_p(fg)$$

determined by taking the expectation of a product of random variables, with respect to the measure p at which we are located. It defines an orthogonal projection

$$\pi_p : R_\Omega \to T_p P$$

where $T_p P$ is identified with the subspace of R_Ω spanned by the scores. The connection defined by projecting $\nabla X(v)$ onto $T_p P$, where v lies in $T_p P$, is called the 1-connection, or exponential connection.

Let us calculate the Christoffel symbols of the exponential connection. Choosing co-ordinates θ we have from the definition

$$\begin{aligned} \Gamma_{ij}^k \frac{\partial}{\partial \theta^k} &= \nabla_i \left(\frac{\partial}{\partial \theta^j} \right) \\ &= \pi_p \left(\frac{\partial}{\partial \theta^i} \left(\frac{\partial \ell}{\partial \theta^j} \right) \right) \\ &= \pi_p \left(\frac{\partial^2 \ell}{\partial \theta^i \partial \theta^j} \right) \end{aligned} \tag{4.5.1}$$

Recall from section 1.5.2 that we can use the inner product on R_Ω to define an inner product g on $T_p P$ by restricting it to the image of $T_p P$ under the score. Relative to the co-ordinate basis this inner product defines a matrix

$$g_{ij}(p) = g_p\left(\frac{\partial}{\partial \theta^i}, \frac{\partial}{\partial \theta^j} \right) = E_p\left(\frac{\partial \ell}{\partial \theta^i}, \frac{\partial \ell}{\partial \theta^j} \right)$$

which is just the Fisher information. If f is a random variable then it is straightforward as in section 1.5.2 to calculate that

$$\pi_p(f) = g^{il}(p) E_p\left(f \frac{\partial \ell}{\partial \theta^i} \right) \frac{\partial}{\partial \theta^l}$$

Applying this to (4.5.1) we find that

$$\Gamma_{jk}^i = g^{il} E_p\left(\frac{\partial^2 \ell}{\partial \theta^j \partial \theta^k} \frac{\partial \ell}{\partial \theta^l} \right) \tag{4.5.2}$$

4.5.1 *The second fundamental form*

We want to show how the general constructions in section 4.4.1 can be applied to yield the second fundamental form of a family. Recall that $\mathcal{P}$ is an affine space and hence has a flat connection. This flat connection arises because the affine structure identifies all the tangent spaces. As $\mathcal{P}$ is a subset of $\mathcal{M}$ the tangent space of $\mathcal{P}$ at p is a subspace of the tangent space to $\mathcal{M}$, that is the space of all random variables. The subspace of R_Ω that is the tangent space to $\mathcal{P}$ at p is the space of all random variables f such that $E_p(f) = 0$. The space $\mathcal{P}$ is an affine space for $R_\Omega/\mathbf{R}.1$ and hence there must be an isomorphism

$$T_p\mathcal{P} \simeq R_\Omega/\mathbf{R}.1$$

This is just the map that sends a random variable f with $E_p(f) = 0$ to the coset $f + \mathbf{R}.1$ with inverse that sends a coset $h + \mathbf{R}.1$ to the random variable $h - E_p(h)$. If ∇ denotes the flat connection on $\mathcal{P}$ with respect to its affine structure we have

$$\nabla(Y)(X) = X(Y(\ell)) - E_p(X(Y(\ell)))$$

This connection can also be constructed using the projection of the flat connection on $\mathcal{M}$ and the projection $R_\Omega \to T_pP$ that sends a random variable f to $f - E_p(f)$.

Using the definition in section 4.4.1 to compute the coefficients of the second fundamental form we see that we recover the same formula as in Chapter 1

$$\begin{aligned}
\alpha_{ij} =& (1-\pi)\nabla_i(\frac{\partial \ell}{\partial \theta^j}) \\
=& (1-\pi)(\frac{\partial^2 \ell}{\partial \theta^i \partial \theta^j} - E_p(\frac{\partial^2 \ell}{\partial \theta^i \partial \theta^j})) \\
=& \frac{\partial^2 \ell}{\partial \theta^i \partial \theta^j} - \sum_{mn} g^{mn} E_p(\frac{\partial^2 \ell}{\partial \theta^i \partial \theta^j}\frac{\partial \ell}{\partial \theta^m})\frac{\partial \ell}{\partial \theta^n} - E_p(\frac{\partial^2 \ell}{\partial \theta^i \partial \theta^j})
\end{aligned}$$

Example 4.5.1 *The* 1-*connection for an exponential family.* Consider the 1-connection for an exponential family P. As we showed in Chapter 1, exponential families are exactly the affine subspaces of $\mathcal{P}$, so that the 1-connection must be the natural flat connection. In particular, canonical parameters or co-ordinates are affine co-ordinates, and must be constant for the 1-connection, i.e. its Christoffel symbols must vanish identically in such co-ordinate systems.

Let us calculate its Christoffel symbols in the canonical co-ordinates using the formulas above. We see immediately that

$$\frac{\partial^2 \ell}{\partial \theta^i \partial \theta^j} = -\frac{\partial^2 K}{\partial \theta^i \partial \theta^j}$$

is a constant random variable. On the other hand the expectation of the score is zero so from equation (4.5.2) we see that the Christoffel symbols vanish.

4.6 Amari's α-connection

Recall from Chapter 3 the α embeddings of Amari (1985)

$$F_\alpha(p) = \begin{cases} \frac{2}{1-\alpha} p^{\frac{1-\alpha}{2}}, & \alpha \neq 1 \\ \log(p), & \alpha = 1 \end{cases}$$

which map P into R_Ω the space of random variables. The tangent space to $F_\alpha(P)$ is spanned by random variables

$$p^{\frac{1-\alpha}{2}} \frac{\partial \ell}{\partial \theta^i}$$

Amari (1985) defines the α-expectation of a random variable by

$$E_p^\alpha(f) = \int_\Omega f p^\alpha \mu$$

Note that if we use the α-expectation to define an inner product on the tangent space to $F_\alpha(P)$ then its matrix with respect to the co-ordinate basis is

$$g_{ij} = E_p^\alpha(p^{\frac{1-\alpha}{2}} \frac{\partial \ell}{\partial \theta^i} p^{\frac{1-\alpha}{2}} \frac{\partial \ell}{\partial \theta^j}) = E_p(\frac{\partial \ell}{\partial \theta^i} \frac{\partial \ell}{\partial \theta^j})$$

which is again the Fisher information matrix.

Using the α-expectation we can define a projection

$$\pi^\alpha : R_\Omega \to T_{F_\alpha(p)}(F_\alpha(P))$$

Composing this with the isomorphism

$$d_pF_\alpha : T_pP \rightarrow T_{F_\alpha(p)}(F_\alpha(P))$$

we can use the methods of section 4.4.1 to define a connection on P which Amari (1985) calls the α-connection ∇^α.

To calculate the Christoffel symbols ${\Gamma^\alpha}_{ij}^m$ of the α-connection we note that

$$\begin{aligned}
\nabla_i^\alpha\left(\frac{\partial}{\partial\theta^j}\right) &= \pi^\alpha\left(\frac{\partial^2}{\partial\theta^i\partial\theta^j}(F^\alpha)\right) \\
&= g^{mn}\,E_p^\alpha\left(\frac{\partial^2}{\partial\theta^i\partial\theta^j}(F^\alpha)\frac{\partial F^\alpha}{\partial\theta^m}\right)\frac{\partial}{\partial\theta^n} \\
&= g^{mn}\left(E_p\left(\frac{\partial^2\ell}{\partial\theta^i\partial\theta^j}\frac{\partial\ell}{\partial\theta^m}\right) + \frac{1-\alpha}{2}E_p\left(\frac{\partial\ell}{\partial\theta^i}\frac{\partial\ell}{\partial\theta^j}\frac{\partial\ell}{\partial\theta^m}\right)\right)\frac{\partial}{\partial\theta^n}
\end{aligned}$$

and hence

$${\Gamma^\alpha}_{ij}^n = g^{nm}\left(E_p\left(\frac{\partial^2\ell}{\partial\theta^i\partial\theta^j}\frac{\partial\ell}{\partial\theta^m}\right) + \frac{1-\alpha}{2}E_p\left(\frac{\partial\ell}{\partial\theta^i}\frac{\partial\ell}{\partial\theta^j}\frac{\partial\ell}{\partial\theta^m}\right)\right)$$

4.7 Connections on the cotangent bundle

The rate of change of 1-forms is a concept in many ways even more important than the rate of change of vector fields. We have already pointed out that for a function f the rate of change of df is a bilinear function of tangent vectors which provides a geometric realisation of the Hessian matrix of f and of second-order Taylor expansion. The expectation of the product of random variables has the interpretation of a bilinear function of tangent vectors, and by attempting to realise it in the form of a rate of change of some 1-form df Amari develops a geometric theory of statistical divergences which we discuss in Chapter 6. Furthermore, as we show below, the natural relationship between the rates of change of vector fields and of 1-forms can be made the basis for understanding the whole family of α-connections introduced by Amari.

The rate of change of differential 1-forms is subject to as much ambiguity as the rate of change of vector fields. For if we were to try a direct approach to defining the rate of change of a 1-form

ω on the plane, say, then we would want to consider some limit of the difference $\omega_{p+tv} - \omega_p$ as t tends to zero. But, just as with tangent vectors, ω_{p+tv} and ω_p are defined on different vector spaces, namely $T_{p+tv}P$ and T_pP, and so they themselves certainly belong to different vector spaces. Once again what this approach lacks is a way of connecting or identifying $T_{p+tv}P$ with T_pP.

As with vector fields, instead of pursuing the idea of connecting tangent spaces we can simply write down the rules that we expect any reasonable definition of a rate of change to obey. The object so defined is called a Koszul connection on the cotangent bundle of P. It is an operation ∇ on 1-forms satisfying the analogue of the rules (4.2.3), namely:

1. $\nabla\omega$ is a function of tangent vectors $v \in T_pP$ which is linear and takes its values in T_p^*P
2. $\nabla(\omega_1 + \omega_2) = \nabla\omega_1 + \nabla\omega_2$ for any two 1-forms ω_1, ω_2
3. $\nabla(f\omega)(v) = df(v)\omega(p) + f(p)\nabla\omega(v)$ for any differentiable function f and every $v \in T_pP$

(4.7.1)

Given a co-ordinate system $(\theta^1, \ldots, \theta^n)$ any 1-form ω can be expanded in the form $\omega = f_1 d\theta^1 + \ldots + f_n d\theta^n$, so that for any connection ∇ on the cotangent bundle we have

$$\nabla\omega(v) = df_1(v)d\theta^1 + f_1(p)\nabla d\theta^1(v) + \ldots \\ + df_n(v)d\theta^n + f_n(p)\nabla d\theta^n(v)$$

It follows that $\nabla\omega$ can be calculated for any 1-form ω once we know the values of

$$\nabla d\theta^1, \ldots, \nabla d\theta^n$$

Since for any tangent vector v $\nabla d\theta^i(v)$ is itself a 1-form we can expand it as

$$\nabla d\theta^i(v) = \bar{A}_1^i(v)d\theta_1 + \ldots + \bar{A}_n^i(v)d\theta^n$$

Since $\nabla d\theta^i$ is linear in v it follows that each of the functions of tangent vectors $\bar{A}_j^i$ are linear in v and hence are themselves 1-forms. They can be expanded

$$\bar{A}_j^i = \sum_{k=1}^{n} \bar{\Gamma}^i_{jk} d\theta^k$$

for some smooth functions $\bar{\Gamma}^i_{jk}$ on the manifold.

The functions $\bar{\Gamma}^i_{jk}$ can be chosen arbitrarily in order to construct the 1-forms $\bar{A}^i_j$, and hence to define a connection ∇ on the cotangent bundle through the formulas above. Notice that

$$\bar{\Gamma}^i_{jk} = \nabla d\theta^i(\frac{\partial}{\partial\theta^j})(\frac{\partial}{\partial\theta^k})$$

Thus $\bar{\Gamma}^i_{jk}$, considered as the jk entries of a matrix of functions $\bar{\Gamma}^i$, can be regarded as a form of Hessian matrix of the co-ordinate function θ^i. The standard definition of the Hessian matrix, as the matrix of second-order partial derivatives of a function with respect to the co-ordinates given, would make this matrix zero. As is usual in differential geometry, what appears as standard is actually just the flat case in affine co-ordinates, and the possibility of varying the geometry means that the Hessian matrix of the co-ordinate functions can in fact be prescribed arbitrarily. The next section lends some justification to these remarks.

As is well known, the second partial derivatives of any smooth function are symmetric, and so therefore is the standard Hessian matrix of a function f. We shall often require that the Hessian matrices ∇df of functions determined by a connection ∇ have the same symmetry property. Considering only the co-ordinate functions $f = \theta^i$, shows that such symmetry requires the matrices of Christoffel symbols $\bar{\Gamma}^i$ to be symmetric, i.e.

$$\bar{\Gamma}^i_{jk} = \bar{\Gamma}^i_{kj}$$

Without going into detail, one can show that symmetry for the co-ordinate functions, and thus of the matrices of Christoffel symbols, is sufficient to guarantee this symmetry for every function f, and thus for every other set of co-ordinate functions. Connections with the property that ∇df is a symmetric bilinear form for every function f are called symmetric connections.

4.8 Dual connections and symmetry

If we regard the choice of a connection on the tangent bundle as determining a particular geometry on a manifold, and the choice of a connection on the cotangent bundle as achieving the same, then it would appear that the choice of either connection should somehow determine the other. Loosely speaking, once the rate of change of vector fields has been specified, the geometry of the manifold has been fixed, and the rate of change of 1-forms should also be determined. This idea can be developed quite explicitly by introducing a particular way of connecting tangent spaces called parallel transport. In the next chapter we will show that a choice of connection on either the tangent or the cotangent bundle determines a parallel transport, and that conversely a parallel transport determines connections on both.

There is another more direct way in which to establish a relationship between connections on the tangent bundle and connections on the cotangent bundle, i.e. the rates of change of vector fields and 1-forms. Each 1-form ω and vector field V together produce a function $\omega(V)$ and it seems reasonable to imagine that changes in the values of this function as its argument moves around the manifold are due solely to changes in ω and V. If V changes to a vector field $V + A$ and ω to a 1-form $\omega + \alpha$ then the changed function would be

$$\Big(\omega+\alpha\Big)(V+A)=\omega(V)\ +\omega(A)+\alpha(V)+\alpha(A)$$

If A and α are small in any sense then $\alpha(A)$ would be small to second-order and should be expected to disappear under any kind of limiting process. These arguments are not meant to be rigorous, but they give a heuristic for the Leibniz rule. They motivate the idea that the rate of change $\nabla V(v)$ of a vector field V under a variation with velocity v and the rate of change $\bar{\nabla}\omega(v)$ of a 1-form should be related to the rate of change of $\omega(V)$ via the following form of Leibniz rule:

$$d\big[\omega(V)\big](v)=\bar{\nabla}\omega(v)\big(V(p)\big)+\omega_p\big(\nabla V(v)\big)$$

for every $v \in T_pP$. This rule can be taken as a fundamental principle relating connections on vector fields and 1-forms, or it can be derived from the theory of parallel transport.

The most important point to realise is that given the connection ∇ the connection $\bar{\nabla}$ is determined uniquely and vice versa. Indeed the Leibniz relation shows how the connection 1-forms of each are determined from the other. Given a co-ordinate system $(\theta^1, \ldots, \theta^n)$ the functions $\theta^i(\partial/\partial\theta^j)$ are either zero or one and are therefore constant functions. Using them in the Leibniz rule gives zero on the left-hand side for every tangent vector v, and hence

$$0 = \bar{\nabla} d\theta^i(v)(\frac{\partial}{\partial\theta^j}) + d\theta^i(\nabla(\frac{\partial}{\partial\theta^j})(v))$$

From the definition of connection forms we have that

$$\bar{A}^i_j(v) = \bar{\nabla} d\theta^i(v)(\frac{\partial}{\partial\theta^j}) \qquad A^i_j(v) = d\theta^i(\nabla\frac{\partial}{\partial\theta^j})(v))$$

so that

$$\bar{A}^i_j(v) = -A^i_j(v)$$

A knowledge of ∇, and particularly its connection 1-forms, therefore allows us to determine $\bar{\nabla}$ completely. Equally well, from $\bar{\nabla}$ and its connection forms we can calculate the connection 1-forms of ∇.

A pair of connections, one on the tangent and one on the cotangent bundle, is called a dual pair when they are related by the Leibniz formula, and it is clear that any connection on either the tangent or the cotangent bundle uniquely determines a connection on the other bundle so that together they form a dual pair. Whenever we have a pair of dual connections we shall usually denote them by the same symbol.

Connections on the cotangent bundle whose duals are flat connections on the tangent bundle are themselves called flat. Notice that, by duality, this is equivalent to the existence of a co-ordinate system $(\theta^1, \ldots, \theta^n)$ for which $(d\theta^1, \ldots, d\theta^n)$ are constant, i.e. $(\nabla d\theta^1, \ldots, \nabla d\theta^n)$ are all zero. These are affine co-ordinates.

Connections on the tangent bundle whose duals are symmetric connections on the cotangent dundle are themselves called symmetric connections. Notice that in any co-ordinate system $(\theta^1, \ldots, \theta^n)$ the Christoffel symbols of a connection ∇ and its dual $\bar{\nabla}$ are related by

$$\Gamma^i_{jk} = -\bar{\Gamma}^i_{jk}$$

Hence the symmetry of a connection ∇ on the tangent bundle can be tested by the symmetry of its own matrices of Christoffel symbols. In particular, flat connections are obviously symmetric. Also the calculation of Christoffel symbols for Amari's 1-connection shows that it too is symmetric.

4.9 Geodesics and the exponential map

Given a connection ∇ our interpretation of ∇df as a Hessian matrix can be justified by finding special co-ordinates $(\theta^1, \ldots, \theta^n)$ for which

$$\nabla df(\frac{\partial}{\partial \theta^i})(\frac{\partial}{\partial \theta^j}) = \frac{\partial^2 f}{\partial \theta^i \partial \theta^j}$$

Applying this to the co-ordinate functions $f = \theta^i$ themselves shows that $\nabla d\theta^i = 0$ for $i = 1, \ldots, n$, so that if the identity is to hold throughout some open subset of a manifold then the co-ordinates must be affine and the connection therefore flat. On the other hand, rather than ask that the identity hold for the one co-ordinate system throughout some open region, we shall show that it is possible to vary the co-ordinates from point to point so that the identity holds at each point in the co-ordinates designated for that point. The geodesics of a connection, which we introduce now, play an essential role in producing such co-ordinate systems.

Given a variation γ we denote its velocity field by $\dot{\gamma}$, i.e. $\dot{\gamma}(t)$ is the velocity of γ at 'time' t, or more precisely the velocity of the variation $s \mapsto \gamma(t+s)$ through $\gamma(t)$. A geodesic for a connection ∇ is a variation γ whose velocity field is constant according to ∇, that is,

$$\nabla\dot{\gamma}(\dot{\gamma}) = 0$$

The curves traced out by geodesics are regarded as the straight lines for the geometry defined by the connection. We denote the function $f \circ \gamma$ by f, suppressing the γ, and we denote its derivative with respect to t by $\dot{f}$. Note that, with this ambiguous use of the notation f, we have $\dot{f} = df(\dot{\gamma})$. Choosing co-ordinates $(\theta^1, \ldots, \theta^n)$

gives

$$\begin{aligned}\dot{\gamma} &= \sum_{i=1}^{n} d\theta^i(\dot{\gamma})\frac{\partial}{\partial\theta^i}\\ &= \sum_{i=1}^{n} \dot{\theta}^i \frac{\partial}{\partial\theta^i}\end{aligned}$$

and so

$$\begin{aligned}\nabla\dot{\gamma}(\dot{\gamma}) &= \nabla\left(\sum_{i=1}^{n}\dot{\theta}^i\frac{\partial}{\partial\theta^i}\right)(\dot{\gamma})\\ &= \sum_{i=1}^{n}\left(d\dot{\theta}^i(\dot{\gamma})\frac{\partial}{\partial\theta^i} + \dot{\theta}^i\nabla\frac{\partial}{\partial\theta^i}(\dot{\gamma})\right)\\ &= \sum_{i=1}^{n}\left(\ddot{\theta}^i\frac{\partial}{\partial\theta^i} + \dot{\theta}^i\sum_{j,k=1}^{n}\Gamma_{ij}^{k}d\theta^j(\dot{\gamma})\frac{\partial}{\partial\theta^k}\right)\\ &= \sum_{i=1}^{n}\left(\ddot{\theta}^i + \sum_{j,k=1}^{n}\Gamma_{jk}^{i}\dot{\theta}^j\dot{\theta}^k\right)\frac{\partial}{\partial\theta^i}\end{aligned}$$

It follows that for $i = 1, \ldots, n$

$$\ddot{\theta}^i + \sum_{j,k=1}^{n}\Gamma_{jk}^{i}\dot{\theta}^j\dot{\theta}^k = 0$$

Notice that if all the Christoffel symbols vanish, i.e. if the co-ordinate system is affine, then we have

$$\ddot{\theta}^i = 0$$

and its solutions are $\theta^i(t) = a^i + v^i t$ for some choices of constants a^i and v^i. These are the uniform motions, and hence parametric forms of straight lines in affine co-ordinates for a flat connection.

Example 4.9.1 *Geodesics on the sphere.* In example 4.4.1 we have calculated the Levi-Civita connection in spherical co-ordinates (θ, ϕ), namely

$$\begin{aligned}\bar{\nabla}\frac{\partial}{\partial\theta} &= -\tan\phi d\phi\frac{\partial}{\partial\theta} + \sin\phi\cos\phi d\theta\frac{\partial}{\partial\phi}\\ \bar{\nabla}\frac{\partial}{\partial\phi} &= -\tan\phi d\theta\frac{\partial}{\partial\theta}\end{aligned}$$

For a variation γ we have

$$\dot{\gamma} = \dot{\theta}\frac{\partial}{\partial\theta} + \dot{\phi}\frac{\partial}{\partial\phi}$$

and the geodesic equation is

$$\begin{aligned}\bar{\nabla}\dot{\gamma}(\dot{\gamma}) &= \ddot{\theta}\frac{\partial}{\partial\theta} + \dot{\theta}\bar{\nabla}(\dot{\gamma}) + \ddot{\phi}\frac{\partial}{\partial\phi} + \dot{\phi}\bar{\nabla}(\dot{\gamma}) \\ &= \ddot{\theta}\frac{\partial}{\partial\theta} + \dot{\theta}\left(\tan\phi\dot{\phi}\frac{\partial}{\partial\theta} + \sin\phi\cos\phi\dot{\theta}\frac{\partial}{\partial\phi}\right) \\ &\quad + \ddot{\phi}\frac{\partial}{\partial\phi} - \dot{\phi}\left(\tan\phi\dot{\theta}\frac{\partial}{\partial\theta}\right) \\ &= 0\end{aligned}$$

This yields the equations

$$\begin{aligned}\ddot{\theta} &= 0 \\ \ddot{\phi} &= -\sin\phi\cos\phi\,\dot{\theta}^2\end{aligned}$$

One can find all of the solutions of these equations by standard methods. However, let us simply note some rather obvious ones. For any constants θ_0, α and β

$$\theta = \theta_0 \quad \text{and} \quad \phi = \alpha + \beta t$$

is a solution. Since θ is the angle of longitude and ϕ the angle of latitude of a point on the sphere, this solution traces out a line of longitude, i.e. constant angle of longitude, at a uniform latitudinal velocity. The lines of longitude form the family of all great circles of the sphere passing through the north and south poles. Given the rotational symmetry of the sphere, and the fact that the construction of the Levi-Civita connection does not seem to violate this symmetry, one might expect that every great circle of the sphere is the trace of a geodesic moving along it with constant angular speed. This is indeed the case, and accounts for all geodesics of the Levi-Civita connection. As supporting evidence we may note that the equator of the sphere is given by $\phi = 0$ and that the variations $\theta = a+bt$, $\phi = 0$ which trace it out are geodesics, since they too are solutions of the geodesic differential equations.

In general the Christoffel symbols of a connection do not vanish and the geodesic equation is a system of second-order differential equations in the co-ordinates as shown above. Standard theorems assert that these systems have a unique solution for each choice of initial values for θ^i and $\dot{\theta}^i$, $i = 1, \ldots, n$, which means a choice of initial velocity v (including location p). Geometrically this says that given a point p and a velocity v in T_pP we can find a unique geodesic through p whose velocity is v, i.e. for which

$$\gamma(0) = p \quad \text{and} \quad \dot{\gamma}(0) = v \in T_pP$$

Let us denote this geodesic by γ_v. The same theorems on systems of differential equations assert that the solution depends smoothly on the initial condition v so that the map $v \mapsto \gamma_v(1)$ is a smooth map. It is called the *exponential map* of the connection, and is denoted

$$\exp_p : T_pP \to P$$

It is the map which assigns to each each tangent vector v the point on the manifold arrived at by following the geodesic with initial velocity v for one unit of time.

It is easy to see that if $t \mapsto \gamma(t)$ is a geodesic so also is $t \mapsto \gamma(at + b)$ for any real numbers a and b. It follows easily that $\gamma_{av}(t) = \gamma_v(at)$ and in particular that $\exp(tv) = \gamma_{tv}(1) = \gamma_v(t)$ so that the curve

$$t \mapsto \exp(tv)$$

is the unique geodesic through p with tangent v. This proves that the derivative map of the exponential map at zero is the identity. Indeed, tv is a variation through 0 in T_pP with velocity v, and its image under $\exp_p$ is γ_v which has initial velocity v. So variations through 0 in T_pP with velocity v map under $\exp_p$ to variations through p in P with velocity v also. Because the derivative map of $\exp_p$ at zero is invertible, being the identity map, the inverse function theorem says that locally the exponential map itself is invertible. If we choose a linear co-ordinate system on T_pP, and compose it with the inverse of the exponential map, this defines a set of co-ordinates on P about p called geodesic co-ordinates at p for the connection ∇. As we shall show, it is for these co-ordinate systems, which vary from point to point, that ∇df is realised as the usual Hessian of f.

First we show that, for geodesic co-ordinates $(\theta^1, \ldots, \theta^n)$ about p , we have $\nabla d\theta^i = 0$ at p for $i = 1, \ldots n$, or dually that $\nabla \partial/\partial\theta^i = 0$. Note that for any co-ordinate system, and any real values $(v^1, \ldots, v^n)$, we have

$$\nabla \left(\sum_{i=1}^{n} v^i \frac{\partial}{\partial\theta^i} \right) \left(\sum_{j=1}^{n} v^j \frac{\partial}{\partial\theta^j} \right) = \sum_{i,j=1}^{n} v^i v^j \nabla\left(\frac{\partial}{\partial\theta^i}\right)\left(\frac{\partial}{\partial\theta^j}\right)$$

Our requirement is equivalent to having the left-hand side of this expression vanish at p for every choice of $(v^1, \ldots, v^n)$. The vector field $\sum_{i=1}^{n} v^i \partial/\partial\theta^i$ gives the velocity field of the variation γ defined by $\theta^i = v^i t$ for $i = 1, \ldots n$, so that we require co-ordinate systems for which these variations are geodesics. Geodesic co-ordinate systems are precisely the ones which meet this requirement.

It is very important to reiterate that geodesic co-ordinates at a point p are not in general affinely related to geodesic co-ordinates at nearby points. This happens only if the Christoffel symbols vanish in the geodesic co-ordinates at all nearby points not just at the point p from which the co-ordinates are emanating, which means that the connection is flat in this neighbourhood of p.

4.9.1 *The second fundamental form and geodesics*

A final remark about geodesics may shed light on the second fundamental form. If $Q \subset P$ is a connection it is natural to consider whether a geodesic which begins at a point of q of Q in a direction tangent to Q actually stays in Q. If this is true for every point of Q then Q is called totally geodesic. The condition for a submanifold to be totally geodesic is that the connection be symmetric and that the second fundamental form of the submanifold should vanish. Notice that we have said nothing so far about projections, so it is not clear how to define the second fundamental form. We do so by introducing a more abstract notion of normal space. If T_qQ is the subspace of T_pP containing the vectors tangent to Q we define N_q to be the vector space quotient T_qP/T_qQ. If X and Y are vector fields on Q then we define the second fundamental form $\alpha(X, Y)$ to be the projection of $\nabla(Y)(X) \in T_qP$ onto N_q. Although this is a slightly abstract definition we can rephrase the condition for α to vanish as follows. The second fundamental form vanishes if, whenever X and Y are vector fields everywhere tangential to Q, we also have $\nabla Y(X)$ tangential to Q.

In the case of an affine space and its flat connection the geodesics are straight lines so the condition for a submanifold to be geodesic is that it contains any line that is tangential to it. This is clearly equivalent to the submanifold being an affine subspace or an open subset of an affine subspace.

4.10 Remarks for Chapter 4

Remark 4.1 For more about the theory of connections see Spivak (1970) or Koboyashi and Nomizu (1963).

4.11 Exercises for Chapter 4

Exercise 4.1 Let $\phi: P \times P \to TP$ be a map which satisfies $\phi(p,q) \in T_pP$. Denote by $\phi_p(q): P \to T_pP$ the map $\phi_p(q) = \phi(p,q)$. If Y is a vector field on P and X a tangent vector at p show that

$$\nabla(Y)(X) = X(Y(\phi_p))$$

defines a symmetric connection with Christoffel symbols

$$\Gamma^i_{jk} = \frac{\partial^2 \phi^i}{\partial \theta^j \theta^k}$$

Exercise 4.2 Let $\phi(p,q) = d_p E_p(\ell(q))$. Show that in this case the connection defined in exercise 1 is Amari's 1-connection.

Exercise 4.3 Let $\phi(p,q) = d_p E_q(\ell(p))$. Show that in this case the connection defined in exercise 1 is Amari's -1-connection.

These exercises are examples of connections defined using *yokes* about which more will be said in Chapter 8.

CHAPTER 5

Curvature

5.1 Introduction

Given a connection, parallel translation means propagating a tangent vector along the trace of a variation in such way that the resulting vector field along the curve is constant according to the connection. In the plane we can do much better than propagate a tangent vector merely along some curve. Indeed, we can propagate any given arrow over the whole plane to form a vector field whose arrows are all parallel translates of each other. In order to propagate a given arrow along a curve in a constant fashion, we need only form its field of parallel translates and restrict that to the curve in question. Hence the name parallel translation.

Parallel translation for any flat connection can be understood in exactly the same way as for the plane. The flat connection defines an affine structure and affine co-ordinate systems. Because of the affine structure any tangent vector v corresponds to a translation, namely the translation vector of the uniform motion corresponding to it. This translation vector defines a uniform motion at every point of the manifold and hence a vector field. These vector fields are taken to be the constant vector fields, or parallel vector fields if we wish to remind ourselves of the situation of the plane. From another point of view, if $(\theta^1, \ldots, \theta^n)$ are affine co-ordinates for a connection ∇ then by definition the vector fields $\partial/\partial\theta^1, \ldots, \partial/\partial\theta^n$ are constant vector fields, and so are their linear combinations. Since any tangent vector is uniquely expressible as a linear combination of the values of these co-ordinate vector fields at its location, every tangent vector arises as the value of some constant vector field. This characterises flat connections as ones with a large enough supply of constant vector fields, indeed of constant co-ordinate vector fields, so that that every tangent vector is the value of a unique constant co-ordinate vector field. Parallel translation along a curve in such a case just amounts to restricting the constant vector field to the curve.

To show that parallel translation need not always be so simple, and in the process to prove that certain connections cannot be flat, let us consider the Levi-Civita connection on the sphere. Let us note that if γ is a geodesic then its own velocity field is constant, so that parallel translation of the initial velocity of a geodesic along the geodesic merely reproduces its own velocity field. Starting at the north pole of the sphere, any tangent vector is the initial velocity of a geodesic which traces out a line of longitude. The velocity field along the line of longitude is therefore the parallel translation of this initial tangent vector. It meets the equator at right angles. The equator itself is traced out by a geodesic, and the parallel translation of a tangent vector at right angles to it is the field of arrows having the same length and remaining at right angles to the equator at every point. We assert this fact without proof, although it follows from the characterisation of Levi-Civita connections to be established in Chapter 6, that the parallel translation of two tangent vectors maintains their inner product, and hence their lengths and the angle between them. The velocity field of the geodesic which traces out the equator is the parallel translation of its velocity at any point, and the translate of a tangent vector at right angles to it must therefore remain at right angles.

Let us now consider the parallel translation of a tangent vector at the north pole, first along the line of longitude to which it is tangent, then along the equator to some different line of longitude, and finally back to the north pole along this second line of longitude. The field of tangent vectors along this triangle points at right angles to the equator, and tangent to the two lines of longitude. Thus the initial tangent vector is directed along the first line of longitude, but the final one, at the end of the journey, is directed along the second (Figure 5.1). The point is that in the affine or flat case, where parallel translation is achieved by forming constant vector fields on the whole space, this difference between initial and final tangent vectors, under parallel translation around a closed curve, simply cannot happen. It follows that the Levi-Civita connection is not flat; it possesses no affine co-ordinate systems.

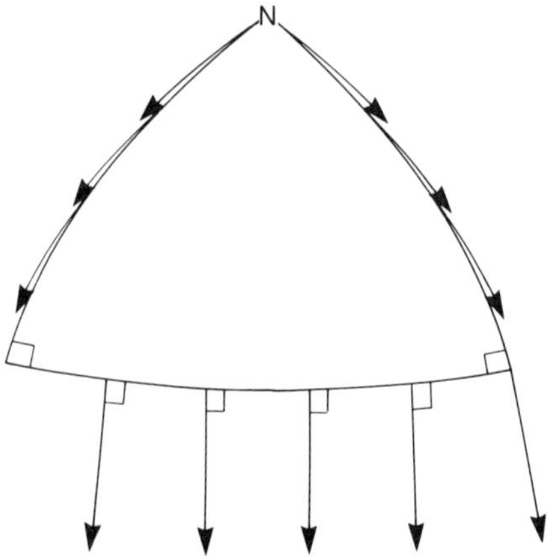

Figure 5.1

One can prove that the Levi-Civita connection is not flat by attacking the equation $\nabla X = 0$ directly to show that it has no constant vector fields. For if a connection has no constant vector fields then it can have no affine co-ordinate systems, the co-ordinate vector fields of affine co-ordinate systems being required to be constant. However this approach works only in special cases, and yields little idea of how to tackle the question of flatness for other connections. Parallel translation provides a more general and geometrical approach to the question of the existence of constant vector fields. In fact it is susceptible to a routine calculus.

Since the parallel translation of a tangent vector along a curve under the Levi-Civita connection produces arrows all of the same length, the initial and final vectors of a parallel translation around a closed curve are related by a rotation. It turns out that the angle of rotation is proportional to the area that the curve encloses. This is quite explicit in the case of the triangles formed of the equator and two lines of longitude, because the area of these triangles is clearly proportional to the length of the equatorial side, which is in turn proportional to the angle between the longitudes. The constant of proportionality is one over the square of the radius of the sphere. For other surfaces the rotation of a tangent vector to its final position under parallel translation around a closed curve is not in direct proportion to the area enclosed, but for small curves it is approximately so. As might be expected, there is a limiting ratio between the angle and area enclosed as the curves are shrunk to the initial point. This limiting ratio is a measure of the curvature of the surace at the point known as the Gaussian curvature.

Rather than deal with general but small closed curves, we can make parallel translation into a computationally effective tool by restricting our attention to 'parallelograms' defined by a pair of tangent vectors located at the same point p. We can use the order of such a pair (v, w) to determine in which of the two possible directions one should proceed around the parallelogram, (v, w) for starting along v first or (w, v) for starting along w first. Given a connection we can determine the parallel translation of any tangent vector u in T_pM around the parallelograms (tv, tw), letting $T_t(v, w)u$ denote the final vector obtained on returning to p. Mindful of the idea that the difference between the final and initial tangent vectors should be proportional to the area of the parallelogram we consider the limit as $t \to 0$ of $T_t(v, w)u - u$ divided by t^2. We denote the limit by $R(v, w)u$. For each (v, w) we can show that $R(v, w)$ is linear as a map $T_pM \to T_pM$, and that R is bilinear and antisymmetric in its dependence on (v, w). For the Levi-Civita connection R is called the Riemann curvature tensor, hence the letter R, and for connections in general it is simply called the curvature tensor, or just the curvature of the connection. For flat connections we must have $R = 0$, and one can show that conversely $R = 0$ implies flatness. This theorem is made into an effective test for flatness by the development of a calculus for the computation of R from a formula for the connection ∇.

5.1.1 *Parallel translation*

Let ∇ be a connection on the tangent bundle of a manifold M, and $\gamma : [a, b] \to M$ a variation starting at $p = \gamma(a)$. For the sake of simplicity we assume that γ is a one-to-one map and that the derivative of γ never vanishes so that its image is a simple curve in M. The parallel translation of a tangent vector v in T_pM under the variation γ is a vector field X along γ which extends v and is constant according to ∇. More precisely it is defined by the conditions

$$X(\gamma(a)) = v \quad \text{and} \quad \nabla X(\dot{\gamma}) = 0$$

If X were the same as $\dot{\gamma}$ then this would be the geodesic equation. However, in this case γ is given and X is to be found.

If $(\theta^1, \ldots, \theta^n)$ is a co-ordinate system then we can expand v and X as

$$v = \sum_{i=1}^{n} v^i \frac{\partial}{\partial\theta^i}(p) \quad \text{and} \quad X = \sum_{i=1}^{n} f^i \frac{\partial}{\partial\theta^i}$$

where $v^1, \ldots, v^n$ are numbers and $f^1, \ldots, f^n$ are functions. Expressing ∇X in this co-ordinate system gives

$$\begin{aligned}\nabla X(\dot\gamma) &= \sum_{i=1}^{n} \left(df^i(\dot\gamma)\frac{\partial}{\partial\theta^i} + f^i \nabla \frac{\partial}{\partial\theta^i} \right) \\ &= \sum_{i=1}^{n} \left(\dot{f}^i \frac{\partial}{\partial\theta^i} + f^i \sum_{j,k=1}^{n} \Gamma_{ij}^k d\theta^j(\dot\gamma)\frac{\partial}{\partial\theta^k} \right)\end{aligned}$$

The equations for parallel transport then become

$$f^i(a) = v^i \quad \text{and} \quad \dot{f}^i + \sum_{j,k=1}^{n} \Gamma_{jk}^i \dot\theta^j f^k = 0$$

This is a system of linear first-order differential equations, and standard theorems guarantee the existence and uniqueness of a solution to this initial-value problem. The uniqueness and linearity also imply that the solution $f^1, \ldots, f^n$ depends linearly on the initial data $v^1, \ldots, v^n$.

Let $q = \gamma(b)$, the end point of γ. From the arguments above, each tangent vector v located at p has a unique parallel translation along γ, and we denote its value at q by $T_\gamma(v)$. This defines a linear map

$$T_\gamma : T_pM \to T_qM$$

also called parallel translation along γ. The context of discussion will always make it clear whether 'parallel translation' means the association of an entire vector field along γ to each tangent vector at its initial point, or just the value of these vector fields at its end point. The map T_γ is invertible, its inverse being given by parallel translation along the reversal of γ. Indeed, the constant vector fields along γ and its reversal are exactly the same, and the same initial-value theory shows that any tangent vector located at q is the value of such a vector field at q. Hence the correspondence between tangent vectors at p and those at q, determined by their being the values of the same vector field constant along γ, is a bijection.

It is possible to recover the connection ∇ from the knowledge of the parallel translation which it determines, i.e. the knowledge of T_γ for each variation γ. In order to calculate $\nabla X(v)$ we first choose a variation γ representing v. Hence $\gamma(0) = p$ is the location of v, and we let T_t denote parallel translation along γ restricted to the interval $[0, t]$. For every t the tangent vector $T_t^{-1}(X(\gamma(t))$ is located at p, so that this is a T_pM-valued function of t. We claim that its derivative at $t = 0$ gives $\nabla X(v)$. To see this let $v_1, \ldots, v_n$ be a basis for T_pM, and let $V_1, \ldots, V_n$ be its parallel translation along γ so that, for each i, V_i is a vector field along γ satisfying

$$\nabla V_i(\dot\gamma) = 0 \quad \text{and} \quad V_i(p) = v_i$$

Because $V_1, \ldots, V_n$ provides a basis for the tangent space of M at each point along γ, the vector field X can be expanded

$$X(\gamma(t)) = f^1(\gamma(t)V_1(\gamma(t)) + \ldots + f^n(\gamma(t))V_n(\gamma(t))$$

for some unique choice of functions $f^1, \ldots, f^n$, so that

$$\begin{aligned}\nabla X(v) &= df^1(v)V_1(p) + f^1(p)\nabla V_1(v) + \ldots \\ &\qquad + df^n(v)V_n(p) + f^n(p)\nabla V_n(v) \\ &= df^1(v)v_1 + \ldots + df^n(v)v_n\end{aligned}$$

On the other hand we have

$$\begin{aligned}T_t^{-1}X(\gamma(t)) &= f^1(\gamma(t))T_t^{-1}V_1(\gamma(t)) + \ldots + f^n(\gamma(t))T_t^{-1}V_n(\gamma(t)) \\ &= f^1(\gamma(t))v_1 + \ldots + f^n(\gamma(t))v_n\end{aligned}$$

which is a T_pM-valued function whose derivative at $t = 0$ is the same as the formula for $\nabla X(v)$ derived above.

The concept of parallel translation determined by a connection works as well for connections on the cotangent bundle as it does for connections on the tangent bundle. Given a connection ∇ on the cotangent bundle, and a variation $\gamma : [a, b] \to M$ which traces out a simple curve, we can consider 1-forms ω along γ which are constant. These are the 1-forms satisfying $\nabla\omega(\dot\gamma) = 0$. Given a co-ordinate system $(\theta^1, \ldots, \theta^n)$ we can expand ω as $\omega = f_1 d\theta^1 + \ldots + f_n d\theta^n$, and thus

$$\begin{aligned}
\nabla\omega(\dot\gamma) &= \nabla\left(\sum_{i=1}^{n} f_i d\theta^i\right)(\dot\gamma) \\
&= \sum_{i=1}^{n} \left(df_i(\dot\gamma) d\theta^i + f_i \nabla d\theta^i(\dot\gamma)\right) \\
&= \sum_{i=1}^{n} \left(df_i(\dot\gamma) d\theta^i + f_i \sum_{j,k=1}^{n} \Gamma^i_{jk} d\theta^j(\dot\gamma) d\theta^k\right) \\
&= \sum_{i=1}^{n} \left(\dot f_i + \sum_{j,k=1}^{n} \Gamma^k_{ji} \dot\theta^j f_k\right) d\theta^i
\end{aligned}$$

The equation for a 1-form constant along γ becomes

$$\dot f_i + \sum_{j,k=1}^{n} \Gamma^k_{ji} \dot\theta^j f_k = 0$$

for $i = 1, \ldots, n$. The standard existence and uniqueness theorems for systems of differential equations once again show that every linear function on the tangent space to a point on γ is the value of some unique 1-form constant along γ. Hence we can set up a linear bijection between $T_p^* M$ and $T_q^* M$ wherein two linear functions are related if they are the values at p and at q of the same 1-form constant along γ. Once again, these bijections are called parallel translation along the variation concerned, and a knowledge of them allows us to reconstruct the connection ∇ which gave rise to them in exactly the same way as for vector fields.

As indicated in Chapter 4, parallel translation provides a geometric basis for the Leibniz principle used to relate connections on the tangent and cotangent bundle. This is because a parallel translation for tangent vectors produces one for cotangent vectors. Indeed, suppose that T_γ is the parallel translation of a tangent vector along a variation γ, determined by some connection ∇. Given a cotangent vector ω located at $p = \gamma(a)$ we can produce one at $q = \gamma(b)$ by translating tangent vectors at q back to p where we can evaluate ω upon them. In other words we define

$$\hat T_\gamma \omega(v) = \omega\left(T_\gamma^{-1}(v)\right)$$

for each v in T_qM. By treating $\hat{T}_\gamma$ as parallel translation of cotangent vectors along γ, we can define a connection on the cotangent bundle by the formula

$$\hat{\nabla}\omega(v) = \frac{d}{dt}\hat{T}_t^{-1}\omega(\gamma(t))_{t=0}$$

where γ represents v and $\hat{T}_t$ is the parallel translation along its restriction to $[0,t]$.

It is easy to show that $\hat{\nabla}$ defined by this formula is the connection determined from ∇ by the Leibniz principle. First note that for any vector field X and 1-form ω

$$\hat{T}_t^{-1}\omega\left(T_t^{-1}X\right)(\gamma(t)) = \omega\left(T_tT_t^{-1}X\right)(\gamma(t)) = \omega(X)(\gamma(t))$$

Both $\hat{T}_t^{-1}\omega(\gamma(t))$ and $T_t^{-1}X(\gamma(t))$ are located at $p = \gamma(0)$, and the Leibniz rule applies to computing the derivative of the evaluation of the former on the latter. Hence

$$\begin{aligned} d\left(\omega(X)\right)(v) &= \frac{d}{dt}\omega(X)(\gamma(t))_{t=0} \\ &= \frac{d}{dt}\hat{T}_t^{-1}\omega\left(T_t^{-1}X\right)(\gamma(t))_{t=0} \\ &= \frac{d}{dt}\hat{T}_t^{-1}\omega(\gamma(t))_{t=0}\left(X(p)\right) \\ &\qquad + \omega_p\left(\frac{d}{dt}T_t^{-1}X(\gamma(t))_{t=0}\right) \\ &= \hat{\nabla}\omega(v)(X(p)) + \omega_p\left(\nabla X(v)\right) \end{aligned}$$

5.1.2 *Differential 2-forms*

A differential 1-form is a function of tangent vectors whose restriction to each tangent space is linear. A differential 2-form is a function of pairs (v, w) of tangent vectors with the same location, whose restriction to each tangent space is bilinear and antisymmetric. For example, if x and y are Cartesian co-ordinates on the plane then

$$A(v,w) = dx(v)dy(w) - dx(w)dy(v)$$

is a differential 2-form. Indeed, if w is fixed then the formula is linear in v, while if v is fixed then it is linear in w. Interchanging v and w reverses the sign of the formula. Thus at any point p, A is a bilinear and antisymmetric function of the tangent vectors located there.

On the plane tangent vectors can be interpreted as arrows, and a pair of arrows based at the same point can be completed into a parallelogram. The order of such a pair (v, w) can be taken to define a direction of circulation around the parallelogram, starting out along v first and returning along w. Hence the pair (w, v) defines the same parallelogram but with the opposite direction of circulation, starting out along w first and returning along v. We call a parallelogram with a specified direction of circulation an *oriented parallelogram* so that an ordered pair of tangent vectors in the plane, located at the same point, determines an oriented parallelogram, and a differential 2-form can be regarded as a function of oriented parallelograms. For example, the 2-form A described above assigns to each oriented parallelogram its area, if its circulation is anticlockwise as determined by the Cartesian co-ordinates, and minus its area if its circulation is clockwise.

In the formula defining A we can replace dx and dy by any other 1-forms θ^1 and θ^2 and the formula will still define a 2-form denoted $\theta^1 \wedge \theta^2$. Thus the formula

$$\theta^1 \wedge \theta^2(v, w) = \theta^1(v)\theta^2(w) - \theta^1(w)\theta^2(v)$$

provides a general procedure, called the wedge product, for producing 2-forms from 1-forms on any manifold. It resembles a multiplication of 1-forms in that it satisfies

$$\begin{aligned} \phi \wedge (\theta^1 + \theta^2) &= \phi \wedge \theta^1 + \phi \wedge \theta^2 \\ \theta^1 \wedge (\lambda\theta^2) &= \lambda(\theta^1 \wedge \theta^2) \\ \theta^2 \wedge \theta^1 &= -\theta^1 \wedge \theta^2 \end{aligned}$$

for any real number λ and 1-forms ϕ, θ^1 and θ^2. Note that the last property says that, unlike multiplication of real numbers, the wedge product anticommutes rather than commutes.

In a sense the wedge product allows us to express all differential 2-forms in terms of differential 1-forms, or more precisely as linear combinations of wedge products. Let $(\phi^1, \ldots, \phi^n)$ be a co-ordinate system on a manifold M, so that $d\phi^1, \ldots, d\phi^n$ are the co-ordinate 1-forms. Every 2-form ω can be expressed

$$\omega = \sum_{i<j}^{n} f_{ij} d\phi^i \wedge d\phi^j$$

where the functions f_{ij} are given by

$$f_{ij} = \omega \left(\frac{\partial}{\partial \phi^i}, \frac{\partial}{\partial \phi^j} \right)$$

Indeed, for any tangent vector v we have

$$v = d\phi^1(v) \frac{\partial}{\partial \phi^1} + \ldots + d\phi^n(v) \frac{\partial}{\partial \phi^n}$$

Using the bilinearity of ω we obtain, for every pair of tangent vectors (v, w),

$$\begin{aligned} \omega(v, w) &= \omega \left(\sum_{i=1}^{n} d\phi^i(v) \frac{\partial}{\partial \phi^i}, \sum_{j=1}^{n} d\phi^j(v) \frac{\partial}{\partial \phi^j} \right) \\ &= \sum_{i,j=1}^{n} \omega \left(\frac{\partial}{\partial \phi^i}, \frac{\partial}{\partial \phi^j} \right) d\phi^i(v) d\phi^j(w) \end{aligned}$$

while the antisymmetry of ω and the definition of $d\phi^i \wedge \phi^j$ give

$$\omega(v, w) = \sum_{i<j}^{n} \omega \left(\frac{\partial}{\partial \phi^i}, \frac{\partial}{\partial \phi^j} \right) d\phi^i \wedge \phi^j(v, w)$$

Let θ be a differential 1-form. For the sake of definiteness, and for motivation, suppose that it is a 1-form on the plane whose value $\theta(v)$ on a tangent vector v approximates the anticlockwise angle through which tangent vectors rotate, under some non-standard parallel translation along the arrow v. In effect, this compares the non-standard against the standard parallel translation in the plane, because the angle of rotation means the angle between the standard and non-standard parallel translates of tangent vectors. Given a pair of tangent vectors (v, w) located at a point p, let us determine the expression for the approximate total angle through which tangent vectors rotate, as they are translated around the oriented parallelogram determined by (v, w) in its given direction of circulation. It will be a sum of four terms, one for each side of the parallelogram. The first term is $\theta(v)$, which we will write as $\theta_p(v)$ to emphasise that the tangent vector is located at p. The next side of the parallelogram to be traversed consists of a copy of w translated (in the usual sense) to the tip of v, i.e. to $p+v$. Hence the angle through which tangent vectors rotate under the unusual parallel translation is $\theta_{p+v}(w)$. The third side to be traversed is parallel to v, and can be considered as a copy of v translated to $p+w$. However, we traverse it in the direction of $-v$, and the approximate rotation of tangent vectors should be $-\theta_{p+w}(v)$. Finally we traverse the arrow w but in the direction $-w$, and so the approximate rotation of tangent vectors should be $-\theta_p(w)$.

Instead of adding the four terms in the order determined by the circulation around the sides of the parallelogram, let us group them according to the pairs of parallel sides. The pair parallel to w contributes the sum $\theta_{p+v}(w) - \theta_p(w)$ while the pair parallel to v contributes $-\theta_{p+w}(v) + \theta_p(v)$. Hence the approximate rotation of tangent vectors as they are translated around the parallelogram is given by

$$\Delta\theta(v, w) = \Big(\theta_{p+v}(w) - \theta_p(w)\Big) - \Big(\theta_{p+w}(v) - \theta_p(v)\Big)$$

This expression has an advantage when, following the introductory discussion of curvature, we scale the parallelogram to (tv, tw) and divide the resulting approximate rotation by t^2. Then we obtain

$$\begin{aligned}\frac{\Delta\theta(tv, tw)}{t^2} &= \frac{\Big(\theta_{p+tv}(tw) - \theta_p(tw)\Big)}{t^2} - \frac{\Big(\theta_{p+tw}(tv) - \theta_p(tv)\Big)}{t^2} \\ &= \frac{\Big(\theta_{p+tv}(w) - \theta_p(w)\Big)}{t} - \frac{\Big(\theta_{p+tw}(v) - \theta_p(v)\Big)}{t}\end{aligned}$$

because of the linearity of θ. This expression appears susceptible of having a limit as $t \to 0$, consisting of some kind of derivative of θ along v, evaluated at w, minus the derivative of θ along w evaluated at v.

In fact, we can say exactly what kind of derivative of the 1-form θ should occur in the limit by emphasising the role of the standard parallel translation in the discussion above. It was used to translate the two tangent vectors v and w into the opposite sides of a parallelogram, upon which θ was then evaluated. If ∇ is the connection defined by this parallel translation, i.e. the natural affine connection in this case, then the limit of the expressions above as $t \to 0$ is precisely the t derivative at zero which expresses ∇ in terms of its parallel translation. It follows that the limit of $\Delta\theta(tv, tw)/t^2$ as $t \to 0$ is given by

$$d\theta(v, w) = \nabla\theta(v)(w) - \nabla\theta(w)(v)$$

It is easy to show that $d\theta$ is a 2-form, and it is called the exterior derivative of θ.

On any manifold, not just the plane, we could write down the same formula for $d\theta$ using any connection. In particular, it would appear that there are a variety of operations of exterior differentiation d, corresponding to the variety of connections ∇ that can be used in the defining formula. From this point of view it may seem surprising that all *symmetric* connections produce the same operation d. We give a simple computational proof of this fact below. However, it has a geometrical basis, namely the parallelogram interpretation from which the formula is derived, and the extent to which this interpretation is possible on a general manifold with connection. Given a pair (v, w) of tangent vectors located at a point p on manifold, and a connection ∇, what does it mean to construct a parallelogram, and how do we interpret translating v along w and w along v to produce one? Tangent vectors cannot be translated along other tangent vectors, but only along the trace of a variation, so we must first choose actual variations with velocities v and w to translate them along. In fact, we shall look for a two-parameter variation $\gamma(s, t)$ such that its velocity at $s = 0$ for each fixed t is the parallel translation of v according to ∇, and its velocity at $t = 0$ for each fixed s is the parallel translation of w.

If the connection is flat then we can use an affine co-ordinate system to construct such a γ. Indeed, by affine changes of co-ordinate if necessary, we can produce an affine co-ordinate system $(\phi^1, \ldots, \phi^n)$ such that $\phi^i(p) = 0$ for $i = 1, \ldots, n$ and $v = (\partial/\partial\phi^1)(p)$ and $w = (\partial/\partial\phi^2)(p)$. A suitable variation γ is then given by the map which assigns to each pair in $[0,1] \times [0,1]$ the point in M which has them as its first two co-ordinates, and its remaining co-ordinates zero. Let us note at this point that it is such 'parallelograms' which allow one to prove the symmetry of the second partial derivatives of functions. The proof rests essentially on the fact that the sum of the changes in a function's values as one traverses the sides of the parallelogram from p back to p is zero. As we have already discussed, the symmetry of the second partial derivatives of functions is equivalent to the symmetry of the bilinear function ∇df for the flat connection ∇ for which the co-ordinates are affine. According to the formula for exterior derivative d, this is equivalent to $d(df) = 0$, which we shall write as $d^2 f = 0$, for every function f.

If the connection ∇ is not flat then in general we cannot construct the 2-parameter variations γ which meet the parallelogram specifications set out above. However, since the formula for the exterior derivative was derived via a limit at the point p, it is necessary to use only infinitesimal parallelograms in the arguments leading to its derivation. Without giving any details, let us say that it is enough to use 2-parameter variations γ whose s and t velocity fields are parallel (have ∇ rate of change zero) only at p itself. One can show that this is equlivalent to the symmetry of ∇, or equivalently to the condition $d^2 f = 0$. This, if you like, is the geometrical significance of symmetry for connections.

Rather than prove the independence of d from the symmetric connection used in its definition by elaborating on these parallelogram arguments, we give a simple calculation proof based on the fundamental condition $d^2 f = 0$. First let us note that, no matter what connection ∇ is used in the formula, d satisfies the first two rules of differential calculus, namely an addition rule and a product or Leibniz rule. The addition rule

$$d(\theta^1 + \theta^2) = d\theta^1 + d\theta^2$$

follows at once from the same property for ∇. The product rule is

$$d(f\theta) = df \wedge \theta + fd\theta$$

Its proof follows from the analogous rule for connections, viz

$$\begin{aligned} d(f\theta)(v,w) &= \nabla(f\theta)(v)(w) - \nabla(f\theta)(w)(v) \\ &= df(v)\theta(w) + f(p)\nabla\theta(v)(w) \\ &\qquad\qquad - df(w)\theta(v) - \ f(p)\nabla\theta(w)(v) \\ &= df \wedge \theta(v,w) + f(p)d\theta(v,w) \end{aligned}$$

These two rules allow for the effective calculation of exterior derivatives, as well as proving that d does not depend on the choice of symmetric connection used in the defining formula. Indeed, given a co-ordinate system $(\phi^1, \ldots, \phi^n)$ we can express any 1-form θ as

$$\theta = f_1 d\phi^1 + \ldots + f_n d\phi^n$$

Since $d^2\phi^i = 0$ for $i = 1, \ldots, n$ in the case of a symmetric connection, we have

$$\begin{aligned} d\theta &= df_1 \wedge d\phi^1 + f_1 d^2\phi^1 + \ldots + \; df_n \wedge d\phi^n + f_n d^2\phi^n \\ &= df_1 \wedge d\phi^1 + \ldots + df_n \wedge d\phi^n \end{aligned}$$

This last expression gives a formula for computing $d\theta$ in terms of co-ordinates. Expressing $df_1, \ldots, df_n$ in terms of co-ordinates gives

$$d\theta = \sum_{i<j}^{n} \left(\frac{\partial f_i}{\partial \phi^j} - \frac{\partial f_j}{\partial \phi^i} \right) d\phi^i \wedge d\phi^j$$

Moreover, this formula in no way depends on the choice of symmetric connection used to define d. This proves that the exterior derivative d is the same for all symmetric connections.

The same formula allows us to prove the third rule of differential calculus, namely the chain rule. The chain rule is about the derivatives of composite functions. Its generalisation to differential forms concerns pullbacks. If $F : M \to N$ is a smooth map between manifolds then 2-forms pull back from N to M under composition with F, or its derivative map, in the same way as functions and 1-forms. A function f on N pulls back to the function $f \circ F$ on M, which we also denote F^*f. A 1-form θ on N pulls back to the 1-form $F^*\theta$ on M, defined by

$$F^*\theta(v) = \theta(TF(v))$$

and a 2-form ω on N pulls back to the 2-form $F^*\omega$ on M defined by

$$F^*\omega(v, w) = \omega(TF(v), TF(w))$$

The general chain rule can be stated in the form

$$dF^* = F^*d$$

in the sense that both sides, applied either to a function or a 1-form, give the same result. For functions the result is almost a tautology, since for any tangent vector v, represented by a variation γ, we have

$$\begin{aligned} dF^*f(v) &= \frac{d}{dt}(f \circ F) \circ \gamma(t)_{t=0} \\ &= \frac{d}{dt} f \circ (F \circ \gamma)(t)_{t=0} \\ &= df(TF(v)) \\ &= d(F^*f)(v) \end{aligned}$$

For the case of 1-forms let us first note that

$$\begin{aligned} F^*(\theta^1 + \theta^2) &= F^*\theta^1 + F^*\theta^2 \\ F^*(\theta^1 \wedge \theta^2) &= F^*\theta^1 \wedge F^*\theta^2 \end{aligned}$$

These relations are trivial to prove. For example in the case of the second,

$$\begin{aligned} F^*(\theta^1 \wedge \theta^2)(v,w) &= (\theta^1 \wedge \theta^2)(TF(v), TF(w)) \\ &= \theta^1(TF(v))\theta^2(TF(w)) - \theta^1(TF(w))\theta^2(TF(v)) \\ &= F^*\theta^1(v)F^*\theta^2(w) - F^*\theta^1(w)F^*\theta^2(v) \\ &= F^*\theta^1 \wedge F^*\theta^2(v,w) \end{aligned}$$

Using these relations, and expanding any 1-form θ in terms of a co-ordinate system, we have

$$\begin{aligned} dF^*\theta &= dF^*(\sum_{i=1}^n f_i d\phi^i) \\ &= d\left(\sum_{i=1}^n F^*f_i dF^*\phi^i\right) \\ &= \left(\sum_{i=1}^n dF^*f_i \wedge dF^*\phi^i\right) \\ &= \left(\sum_{i=1}^n F^*df_i \wedge F^*d\phi^i\right) \\ &= F^*\left(\sum_{i=1}^n df_i \wedge d\phi^i\right) \\ &= F^*d\theta \end{aligned}$$

This proves the chain rule for 1-forms.

It should be noted that the concept of differential 1- and 2-forms extends more generally to differential k-forms for any natural number k, these being functions of k tangent vectors at the same location, whose restriction to any one tangent space is multilinear and antisymmetric. The notion of their sum, wedge product and pull-back under a smooth map can all be defined, as well as their exterior derivative. The calculus of differential forms consists of the sum, product and chain rule for the exterior derivative in this general case, along with the concept of the integral of a differential k-form along a k-parameter variation and the resulting fundamental theorem of calculus, and the Poincaré lemma on the existence of primitives, i.e. solutions of the equation $d\theta = \omega$ for given ω. These theorems provide a uniform and dimension-free treatment of classical vector analysis, extending the concepts of divergence, gradient, curl, various integrals of vector fields and Gauss's, Green's and Stokes' Theorems. We have not covered all of these these topics, since it appears to us to move too far from what may be relevant to statistics. However, the Poincaré lemma will be of particular importance.

Since for any function f we have $d(df) = 0$, a necessary condition for there to be a solution f to the equation $df = \theta$, where θ is a given 1-form, is that $d\theta = 0$. The Poincaré lemma asserts that on a region of the manifold having a co-ordinate system whose co-ordinates form a convex set, or more generally a star-shaped set, this condition is sufficient to guarantee a solution of the equation defined on that region. Recall that a star-shaped region is one which has a point p with the property that for every other point q in the region the line segment joining p to q is in the region. This is, of coures weaker than being convex. The Poincaré lemma is formulated for differential forms of arbitrary degree. In the case which we are discussing the argument is much simpler than the general theorem, and corresponds to the well known exactness theory for ordinary differential equations. Indeed, if we express θ in co-ordinates $(\phi^1, \ldots, \phi^n)$, as $f_1 d\phi^1 + \ldots f_n d\phi^n$ then $d\theta = 0$ is equivalent to

$$\frac{\partial f_i}{\partial \phi^j} = \frac{\partial f_j}{\partial \phi^i}$$

As is well known, these conditions allow us to find a function f satisfying $\partial f / \partial \phi^i = f_i$ for $i = 1, \ldots n$ over suitable domains. This

f satisfies $df = \theta$.

5.1.3 *Curvature*

Just as a function f produces a 1-form df, so a connection ∇ and a vector field X may be regarded as producing a 1-form ∇X with tangent vector values as opposed to real values. That is, ∇X is a function of tangent vectors v whose restriction to any tangent space is linear, but whose value $\nabla X(v)$ is a tangent vector with the same location as v rather than a real number. We shall repeat the construction of the exterior derivative for such 1-forms, denoting it by ∇ also, and show that $\nabla(\nabla X)$ at any point p describes the infinitesimal parallel translation of $X(p)$ around parallelograms. In particular, while $d^2 = 0$ is fundamental to the exterior derivative of ordinary differential forms, it is not usually the case that $\nabla^2 = 0$ for the exterior derivative of vector-valued forms arising from a connection ∇. In effect ∇^2 becomes the definition of the curvature of ∇, and the calculus of these types of differential forms provides an effective way of computing it.

Given a pair of tangent vectors (v, w) located at a point p let us choose a flat connection $\bar{\nabla}$ with which to form parallelograms from them. Via the affine structure of the flat connection we can identify tangent vectors with translations, and make sense of the uniform variations $p + tv$ and $p + tw$. Let $\bar{T}_v w$ be the parallel translate of w along $p + tv$ as far as $p + v$, and similarly let $\bar{T}_w v$ be the parallel translate of v along $p + tw$ as far as $p + w$ as in Figure 5.2. They effectively form the sides of the parallelogram opposite w and v respectively. For any ordinary differential 1-form θ we introduce the exterior derivative $d\theta$ as a limit of the differences $\theta(\bar{T}_{tv} w) - \theta(w)$ and $\theta(\bar{T}_{tw} v) - \theta(v)$ divided by t. If Θ is a vector-valued form these differences no longer make sense, because $\Theta(\bar{T}_{tv} w)$ is a value located at $p + tv$ while $\Theta(w)$ is located at p. We need a connection in order to make sense of these differences. Let T_{tv} be the parallel translation along $p + tv$ according to the given connection ∇, and likewise T_{tw} be the parallel translation along $p + tw$. We define

$$\Delta\Theta(v, w) = \frac{(T_{tv}^{-1}\Theta(\bar{T}_{tv} tw) - \Theta(tw))}{t^2} - \frac{(T_{tw}^{-1}\Theta(\bar{T}_{tw} tv) - \Theta(tv))}{t^2}$$

Then we define

$$\nabla\Theta(v, w) = \lim_{t \to 0} \Delta\Theta(v, w)$$

where the limit exists. In the event that it exists for every pair (v, w) and defines a vector-valued 2-form we call it the exterior *covariant* derivative of Θ.

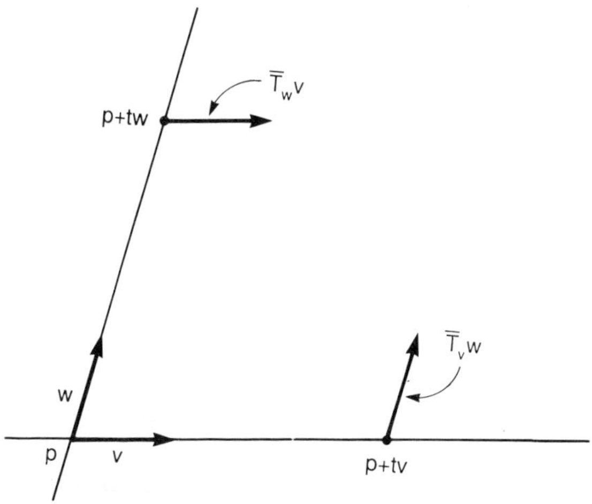

Figure 5.2

It is quite obvious that, where the exterior covariant derivatives each of two vector-valued 1-forms exist, we have addition rule:

$$\nabla(\Theta_1 + \Theta_2) = \nabla\Theta_1 + \nabla\Theta_2$$

More importantly, the following form of Leibniz rule holds. If X is a vector field and θ an ordinary 1-form then we can multiply them to produce a vector-valued 1-form by defining

$$(\theta X)(v) = \theta(v)X(p)$$

where p is the location of v. Often this product is written $\theta \otimes X$ and called the tensor product of θ and X, but there is no necessity for this extra complication here. If Θ is such a product then we have

$$\begin{aligned}
&\lim_{t\to 0}\frac{(T_{tv}^{-1}\Theta(\bar{T}_{tv}tw)-\Theta(tw))}{t^2}\\
&=\lim_{t\to 0}\frac{(\theta(\bar{T}_{tv}w)T_{tv}^{-1}X(p+tv)-\theta(w)X(p))}{t}\\
&=\lim_{t\to 0}\frac{(\theta(\bar{T}_{tv}w)-\theta(w))T_{tv}^{-1}X(p+tv)}{t}\\
&\qquad+\lim_{t\to 0}\frac{(\theta(w)(T_{tv}^{-1}X(p+tv)-X(p))}{t}\\
&=\bar{\nabla}\theta(v)(w)X(p)+\theta(w)\nabla X(v)
\end{aligned}$$

Treating the second term in the definition of $\Delta\Theta(v,w)$ in the same way we conclude that

$$\begin{aligned}
\nabla(\theta X)(v,w) = \bar{\nabla}\theta(v)(w)X(p) + {}& \theta(w)\nabla X(v)\\
&- \bar{\nabla}\theta(w)(v)X(p) - \theta(v)\nabla X(w)
\end{aligned}$$

By analogy with the case of ordinary 1-forms, we define the wedge product of an ordinary 1-form θ with a vector-valued 1-form Θ by

$$\theta\wedge\Theta(v,w)=\theta(v)\Theta(w)-\theta(w)\Theta(v)$$

It follows that

$$\nabla(\theta X)(v,w)=d\theta(v,w)X(p)-\theta\wedge\nabla X(v,w)$$

This argument shows that for any smooth 1-form θ and vector field X the exterior covariant derivative of the vector-valued 1-form θX exists and can be computed by the rule

$$\nabla(\theta X)=d\theta X-\theta\wedge\nabla X$$

However, every smooth vector-valued 1-form Θ can be expressed as a sum of terms of this form by expanding it in a co-ordinate system. Given a co-ordinate system $(\phi^1,\ldots,\phi^n)$ we can expand the values $\Theta(v)$ in terms of the basis $(\partial/\partial\phi^1)(p),\ldots,(\partial/\partial\phi^n)(p)$ to give

$$\Theta(v)=A^1(v)\frac{\partial}{\partial\phi^1}(p)+\ldots+A^n(v)\frac{\partial}{\partial\phi^n}(p)$$

Since Θ is linear in v on any tangent space, so are the A^i, and hence they are 1-forms. The smoothness of Θ means that they are smooth, and we have the expansion

$$\Theta = A^1 \frac{\partial}{\partial \phi^1} + \ldots + A^n \frac{\partial}{\partial \phi^n}$$

Hence the addition and Leibniz rules show that any smooth vector-valued 1-form has an exterior covariant derivative, and they provide an effective calculus for its computation.

We can apply these ideas in particular to the vector-valued 1-forms ∇X, the covariant derivatives of vector fields. In fact, recall that in order to specify a connection ∇, or to calculate with it, one expands the vector valued 1 forms $\nabla(\partial/\partial\phi^i)$ as above, i.e. in the form

$$\nabla \frac{\partial}{\partial \phi^i} = A_i^1 \frac{\partial}{\partial \phi^1} + \ldots + A_i^n \frac{\partial}{\partial \phi^n}$$

where the A_i^k are called the *connection 1-forms* of ∇. Accordingly we obtain the general formulas

$$\begin{aligned}
\nabla(\nabla \frac{\partial}{\partial \phi^i}) &= \sum_{k=1}^{n} \left(dA_i^k \frac{\partial}{\partial \phi^k} - A_i^k \wedge \nabla \frac{\partial}{\partial \phi^k} \right) \\
&= \sum_{k=0}^{n} \left(dA_i^k \frac{\partial}{\partial \phi^k} - \sum_{l=0}^{n} A_i^k \wedge A_k^l \frac{\partial}{\partial \phi^l} \right) \\
&= \sum_{k=0}^{n} \left(dA_i^k - \sum_{l=0}^{n} A_i^l \wedge A_l^k \right) \frac{\partial}{\partial \phi^k}
\end{aligned}$$

However, it is usually easier to calculate $\nabla^2(\partial/\partial\phi^i)$ directly from the rules of the calculus than to remember these formulae.

Example 5.1.1 *The Levi-Civita connection on the sphere.* If ∇ is the Levi-Civita connection on the sphere then for the spherical co-ordinates (θ, ϕ) we have determined that

$$\begin{aligned}
\nabla \frac{\partial}{\partial \theta} &= \tan\phi d\phi \frac{\partial}{\partial \theta} + \sin\phi\cos\phi d\theta \frac{\partial}{\partial \phi} \\
\nabla \frac{\partial}{\partial \phi} &= -\tan\phi d\theta \frac{\partial}{\partial \theta}
\end{aligned}$$

Either from this, or directly from the derivation of the Levi-Civita connection, we can show that

$$\nabla\left(\frac{1}{\cos\phi}\frac{\partial}{\partial\theta}\right) = \sin\phi d\theta\frac{\partial}{\partial\phi}$$

$$\nabla\frac{\partial}{\partial\phi} = -\sin\phi d\theta\left(\frac{1}{\cos\phi}\frac{\partial}{\partial\theta}\right)$$

It follows that

$$\begin{aligned}\nabla^2\left(\frac{1}{\cos\phi}\frac{\partial}{\partial\theta}\right) &= \cos\phi d\phi\wedge d\theta\frac{\partial}{\partial\phi}\\ &\qquad -\sin\phi d\theta\wedge\left(-\sin\phi d\theta\left(\frac{1}{\cos\phi}\frac{\partial}{\partial\theta}\right)\right)\\ &= \cos\phi d\phi\wedge d\theta\frac{\partial}{\partial\phi}\end{aligned}$$

where the second term has vanished because $d\theta\wedge d\theta = 0$. Similarly we can show that

$$\nabla^2\frac{\partial}{\partial\phi} = -\cos\phi d\phi\wedge d\theta\left(\frac{1}{\cos\phi}\frac{\partial}{\partial\theta}\right)$$

While we have calculated

$$\nabla^2\left(\frac{1}{\cos\phi}\frac{\partial}{\partial\theta}\right)$$

because it was easier to do so, it may be that what we really wanted to know was $\nabla^2(\partial/\partial\theta)$. It turns out that

$$\nabla^2\left(\frac{1}{\cos\phi}\frac{\partial}{\partial\theta}\right) = \frac{1}{\cos\phi}\nabla^2\frac{\partial}{\partial\theta}$$

from which we deduce that

$$\nabla^2\frac{\partial}{\partial\theta} = \cos^2\phi d\phi\wedge d\theta\frac{\partial}{\partial\phi}$$

More generally we can prove that, for any function f and vector field X, we have

$$\begin{aligned}\nabla^2(fX) &= \nabla(dfX + f\nabla X)\\ &= d^2fX - df\wedge\nabla X + df\wedge\nabla X + f\nabla^2X\\ &= f\nabla^2X\end{aligned}$$

We would have expected a formula for $\nabla^2(fX)$ to involve df and thus to depend on the derivatives of f rather then just its values. The fact that it does not reflects the fact that $\nabla^2 X$ computes the sum of the changes in the values of the vector field X under parallel translation as we traverse the edges of a parallelogram, in exactly the same way as $d^2 f$ computes the sum of the changes in the values of the function f.

On account of the linearity of ∇^2 with respect to functions we can calculate $\nabla^2 X$ for any vector field X from a knowledge of the basic cases $\nabla^2(\partial/\partial\phi^i), i = 1, \ldots n$, by expanding X in the form $f_1(\partial/\partial\phi^1) + \ldots + f_n(\partial/\partial\phi^n)$. In fact, it follows that in general $\nabla^2 X(v, w)$ has the form

$$\nabla^2 X(v, w) = R(v, w)X$$

where $R(v, w)$ is a linear map $T_p M \to T_p M$. Computing $\nabla^2(\partial/\partial\phi^i)$ and expanding the result in terms of $\partial/\partial\phi^j$ amounts simply to determining the matrix of the linear transformation R with respect to the basis $\partial/\partial\phi^i$, where we have suppressed the dependence on the pair (v, w). If we define the 2-forms R_i^k by

$$R(v, w)\frac{\partial}{\partial\phi^i} = \sum_{k=1}^{n} R_i^k(v, w)\frac{\partial}{\partial\phi^k}$$

then according to the calculation above we have

$$R_i^k = dA_i^k - \sum_{l=0}^{n} A_i^l \wedge A_l^k$$

R itself is a 2-form, since it is a function of pairs of tangent vectors located at the same point, however its values are linear maps from the tangent space at that location to itself. In short, we say that R is an operator-valued 2-form.

The operator-valued 2-form R is called the *curvature* of the connection ∇. Its matrix R_i^k is usually easier to determine by calculating $\nabla^2(\partial/\partial\phi^i)$ directly via the exterior differential calculus, and expressing the results in terms of $\partial/\partial\phi^k$, rather than to remember the formulas above.

Example 5.1.2 *The curvature of the Levi-Civita connection on the sphere.* In the example above we have calculated

$$\nabla^2 \frac{\partial}{\partial \theta} = \cos^2 \phi d\phi \wedge d\theta \frac{\partial}{\partial \phi}$$

$$\nabla^2 \frac{\partial}{\partial \phi} = -d\phi \wedge d\theta \frac{\partial}{\partial \theta}$$

so that the matrix of R with respect to the basis $\partial/\partial\theta, \partial/\partial\phi$ is

$$R(\frac{\partial}{\partial \theta}, \frac{\partial}{\partial \phi}) = \begin{pmatrix} 0 & -1 \\ \cos^2 \phi & 0 \end{pmatrix}$$

5.1.4 *Vanishing curvature and flatness*

If a connection ∇ is flat then by definition it has affine co-ordinate systems, co-ordinate systems $(\phi^1, \ldots, \phi^n)$ for which the co-ordinate vector fields $\partial/\partial\phi^1, \ldots, \partial/\partial\phi^n$ are constant. It follows that $\nabla^2(\partial/\partial\phi^i) = 0$ for each $i = 1, \ldots, n$ so that $R = 0$. In particular, since for the Levi-Civita connection of the sphere, the matrix of R is non-zero in the spherical co-ordinate system, we can say that R itself is non-zero, and it follows that the Levi-Civita connection for the sphere is not flat. This example shows how the curvature of a connection can be routinely computed using (exterior) differential calculus, and how one can prove by the non vanishing of its curvature that a connection is not flat.

It turns out that one can prove the converse, that when $R = 0$ then the connection is flat. This is something like the Poincaré lemma. What we need to show is that there are enough solutions of the equation $\nabla\theta = 0$ for 1-forms θ. By 'enough'we mean that there are solutions $(\theta^1, \ldots, \theta^n)$ whose values at some point p are a basis for the tangent space there; in effect an n- dimensional family of solutions, where n is the dimension of the manifold M. Since these 1-forms are constant their values at any other point q are given by parallel translation of their values at p to q along any variation. Since parallel translation is a linear isomorphism the values of $(\theta^1, \ldots, \theta^n)$ at q span the tangent space at q also, and hence at every point.

These solutions provide an affine co-ordinate system for the connection. Note that for a constant 1-form θ

$$d\theta(v, w) = \nabla\theta(v)(w) - \nabla\theta(w)(v) = 0$$

Hence by the Poincaré lemma any such constant 1-form is exact, i.e has the form $d\psi$ for some function ψ. Choosing functions $(\psi^1, \ldots, \psi^n)$ such that $d\psi^i = \theta^i, i = 1, \ldots n$ for the n dimensional family of constant 1-forms, we obtain an affine co-ordinate system. Indeed, $d\psi^1, \ldots d\psi^n$ have linearly indepent values at each point, and are therefore a local co-ordinate system, and they are constant, so they are an affine co-ordinate system.

The vanishing of curvature is therefore a criterion for the existence of special systems of co-ordinates. Kass (1984) has considered the statistical significance of these special co-ordinates for various values of the parameter α in the α-connection. In Chapter 9 we shall consider a related problem of when special co-ordinate systems exist at one particular point.

5.2 Remarks for Chapter 5

Remark 5.1 For references on curvature see see Spivak (1970) or Koboyashi and Nomizu (1963).

Remark 5.2 Curvature has profound importance in differential geometry because of its relationship with topology. For instance the Gauss–Bonnet theorem says that the integral of the curvature over a closed surface is 2π times the Euler class of the surface. The Euler class of the surface is $2 - 2g$ where g is the genus or number of holes in the surface. The interested reader should consult Klingenberg (1978). There seems at present no application of these global topological ideas in statistics.

5.3 Exercises for Chapter 5

Exercise 5.1 Calculate the curvature of the cylindrical connection defined in Example 4.2.1.

Exercise 5.2 Consider the circle $S^1 = \{(x, y) \mid x^2 + y^2 = 1\}$ with the connection induced by orthogonal projection as in Chapter 4. If we identify the tangent space at $(1, 0)$ in the obvious way, and we identify $\mathbf{R}^2$ with the complex numbers, show that the exponential map is the usual complex exponential function.

Exercise 5.3 Calculate the curvature of the 0-connection on the space $\mathcal{N}$ of all normal densities.

CHAPTER 6

Information metrics and statistical divergences

6.1 Introduction

In Chapter 1 we introduced the expected Fisher information of a parametric family, which is the matrix of functions

$$E_p(\frac{\partial \ell}{\partial \theta^i}\frac{\partial \ell}{\partial \theta^j})$$

The earliest ideas about a relationship between statistics and differential geometry stem from the interpretation of the Fisher information as a Riemannian metric by Rao (1945). A Riemannian metric on a manifold is an inner product given on each tangent space. This is not the only Riemannian metric that can be defined on a statistical manifold. Barndorff-Nielsen (1986a) has introduced the *observed information*, which also acts as a Riemannian metric, and a whole theory of observed geometry which includes a general theory for producing such Riemannian metrics, called yokes. This theory of yokes also can also be used to define the expected Fisher information.

In classical differential geometry a Riemannian metric introduces a definition of distance between the points of a manifold, hence the name 'metric'. However, it does not seem that this particular concept of distance between points is of any real statistical significance, and Riemannian metrics play quite a different role instead. A second role for Riemannian metrics is inspired by the general theory of relativity, where the set of events in the cosmos is postulated as a four-dimensional manifold, the space-time manifold, and there is a quadratic function on each tangent space that models the equation of a light ray. The quadratic function is the direct analogue of a Riemannian metric and, as in statistics, there were great efforts to find significance in the analogous concept of distance between events that it determines. However, this is not the role in general relativity that the Riemannian, or rather Lorentzian, metric actually plays. (A Lorentzian metric defines a non-degenerate metric of signature $(+,-,-,-)$ on each tangent space rather than a positive definite one.) Instead it is used to determine a particular Koszul connection, namely one for which the rate of change of the metric is constant. All of the laws of physics are formulated in terms of the rates of change of certain vector fields, 1-forms and their analogues, so that the interpretation of these laws, and their consequences, depend on a choice of Koszul connection. The idea is that the Lorentzian metric determines not merely the equation of light rays, but in some sense the general flow of energy in space-time. The principle that the laws of physics should be such that the speed of light appears the same to all observers is generalised to the principle that the flow of energy should appear the same. The mathematical formulation of this principle is simply that the Koszul connection used to determine the rates of change of vector fields and 1-forms should be symmetric, and chosen so that the rate of change of the Lorentzian metric is zero. It is a fundamental theorem of Riemannian geometry that there is a unique such connection, called the Levi-Civita connection. In the case of a statistical manifold with the Fisher information metric this is Amari's 0-connection which we discussed in Chapter 4. One might have hoped that the Fisher or observed information were statistical quantities that needed to be considered constant near any point of the statistical family; a sort of unbiasedness principle which would favour the choice of Levi-Civita connection for the metric. However, this does not appear to be of significance either.

A more important role for the Riemannian metric, taken up by Amari in statistics, is to determine a duality, i.e. an exact correspondence between vector fields and 1-forms. Any inner product on a vector space achieves an exact correspondence between vectors in the space and linear functions on it, and a Riemmanian metric therefore achieves such a correspondence on each tangent space. Functions with tangent vector values thereby correspond to functions with cotangent vector values and thus vector fields with

1-forms. By this correspondence a rate of change of vector fields can also be regarded as the rate of change of 1-forms and vice versa. In particular, if ∇ is a Koszul connection on the tangent bundle of a manifold, and ∇ also denotes the natural connection which it defines on the cotangent bundle, then a Riemmannian metric will allow us to regard this latter as a connection $\hat{\nabla}$ back on the tangent bundle also. Amari calls $\hat{\nabla}$ the dual of ∇. A connection such that $\nabla = \hat{\nabla}$ is called self dual, and self duality of a connection turns out to be equivalent to the condition that the rate of change of the Riemmannian metric defining the duality be constant. Thus symmetry and self duality are yet another characterisation of the Levi-Civita connection. If ∇ and $\hat{\nabla}$ differ, as they do for example in the case of the exponential connection, then the fact that any convex combination of the two connections is still a connection results in a whole 1-parameter family of connections

$$\left(\frac{1-\alpha}{2}\right)\hat{\nabla} + \left(\frac{1+\alpha}{2}\right)\nabla$$

This gives an alternative definition of Amari's α-connections.

In the case of statistics there is yet a different role for a Riemannian metric, such as the Fisher information or the observed information of Barndorff-Nielsen. Statistical manifolds come already equipped with connections, such as the exponential or 1-connection and its dual. Some of Amari's work in this area can be interpreted as showing, in the case of a flat connection, that the Riemannian metric should be interpreted as a geometric second derivative of a statistical divergence. If g is a Riemannian metric, so that $g(v, w)$ is the inner product between tangent vectors v and w, then we can regard it as a function of single tangent vectors, through its first argument, whose value on a tangent vector v is the linear function $g(v, -)$ in its second argument. In other words, a Riemannian metric can be regarded as a 1-form whose values are 1-forms. This is exactly the kind of object that would result from the rate of change of a 1-form, and Amari's idea is to express g in this way, i.e. to solve the equation

$$\nabla\theta = g$$

where ∇ is the given connection, and θ is some 1-form. If θ exists it has to be closed so that, at least locally, $\theta = d\psi$ for some function ψ. Both θ and ψ can be made unique by specifying initial conditions, and we let ψ_p be the solution for which $\psi_p(p) = 0$ and $d\psi_p = 0$. It turns out that $\hat{\nabla}$ must be symmetric and ∇ must be flat if we are to be able to find such solutions for every point p.

If ∇ is the exponential connection then ψ_p reproduces the cumulant generating function K in the case of an exponential family with p as origin. Amari's theory therefore shows that the cumulant generating function is part of the intrinsic geometry of an exponential family. On the other hand, if ∇ is the dual of the exponential connection then $\psi_p(q)$ is the Kullback–Leibler information of q relative to p, which is also therefore an intrinsic geometric quantity. These results are in the nature of *theorema egregia* for statistical families. The Gaussian curvature of a surface was originally understood in terms of the ratio of the area swept out by the tip of a unit normal to the surface to the area swept out by its base on the surface. As such it appeared to depend on the way in which the surface curved around in space. Gauss, in his *theorem egregium*, showed that the curvature could be determined by measurements relying only on distances within the surface itself, and depending in no way on the embedding of the surface in space, so that surfaces like a plane and a cylinder which could be mapped onto each other in a way that preserved the distances between points, e.g. by unrolling the cylinder, must have the same Gaussian curvature. Amari's theory shows likewise that the cumulant generating function and the Kullback–Leibler information is determined by geometric data pertaining only to the statistical family itself.

By the duality determined by a Riemannian metric g, the 1-form θ which satisfies $\nabla\theta = g$ corresponds to a vector field X_θ. It is easy to show that the integral variations of X_θ, i.e. variations whose tangent vector at each point is give by the value of X_θ there, must trace out geodesics of $\hat{\nabla}$. This leads immediately to Amari's characterisation of points on a surface which minimise the divergence from some given point p, as those lying on geodesics through p which are orthogonal to the surface. This generalises the standard Euclidean result that the point on a surface of minimum distance from a given point p lies at the foot of a perpendicular to the surface from p.

6.2 Riemannian metrics

A Riemannian metric is a positive definite bilinear function on each tangent space which plays the role of an inner product on each tangent space. For example if P is a submanifold of $\mathbf{R}^n$ then, as explained in Chapter 3, any tangent vector of P is included as a tangent vector of $\mathbf{R}^n$ itself, and is identified as a vector in $\mathbf{R}^n$ through the affine structure of $\mathbf{R}^n$. Given any two tangent vectors v and w in T_pP we can define $g(v, w)$ to be their inner product considered as elements of $\mathbf{R}^n$. A manifold with a Riemannian metric is called a Riemannian manifold.

The concept of a Riemannian metric was introduced into geometry in order to provide a notion of speed in the manifold, and of the distance between two points. Any inner product on a vector space determines a notion of length of vectors and of angles between them. If g is an inner product on a vector space V then the length of any vector is v defined to be

$$\sqrt{(g(v,v))}$$

The length of an abstract vector is really just a measure of its size, or notion of its magnitude. Having introduced the concept of the length of a tangent vector, we can recover the notion of distance. Let $\gamma\colon [0,1] \to M$ be a smooth variation from a point p to a point q. The distance traversed by the variation should be the integral of its speed with respect to time. Thus we can define the distance traversed by γ to be the integral of the length of its tangent vector, namely

$$\text{length}(\gamma) = \int_0^1 \sqrt{(g_{\gamma(t)}(\gamma'(t), \gamma'(t))}dt$$

It is straightforward to check, via the change of variables formula, that this only depends on the image of the map γ, and is not changed by reparametrisations of it. Knowing the distance traversed by any variation from p to q, we can define the distance between them to be

$$d(x,y) = \inf_{\{\gamma|\gamma(0)=p,\gamma(1)=q\}} \{\text{length}(\gamma)\}$$

the infimum being over all variations from p to q. This satisfies the axioms for a metric, making the manifold into a metric space, viz.

1. $d(x, y) \geq 0$ and $d(x, y) = 0$ if and only if $x = y$
2. $d(x, y) = d(y, x)$
3. $d(x, y) \leq d(x, z) + d(z, y)$.

If $\theta^1, \ldots, \theta^n$ is any co-ordinate system on a manifold M then any tangent vector v can be expanded

$$v = \sum_{i=1}^{n} d\theta^i(v) \frac{\partial}{\partial \theta^i}$$

Given any Riemannian metric g, its bilinearity along with the expansion above gives

$$g(v, w) = \sum_{i,j=1}^{n} g\left(\frac{\partial}{\partial \theta^i}, \frac{\partial}{\partial \theta^j}\right) d\theta^i(v) d\theta^j(w)$$

This is the form which a Riemannian metric takes when expressed in co-ordinates. The functions

$$g_{ij} = g\left(\frac{\partial}{\partial \theta^i}, \frac{\partial}{\partial \theta^j}\right)$$

are called the co-efficients of the metric in the chosen co-ordinates. In a more co-ordinate dependent treatment these functions are called the metric tensor. If γ is any variation then recall that the co-ordinate functions $\theta^i(\gamma(t))$ are often abbreviated $\theta^i(t)$, and the expansion of its tangent vector $\gamma'(t)$ abbreviated as

$$\sum_{i=0}^{n} \dot{\theta}^i(t) \frac{\partial}{\partial \theta^i}$$

Therefore in co-ordinates the formula for the length of the curve traced out by the variation becomes

$$\text{length}(\gamma) = \int_0^1 \sqrt{\sum_{i,j=1}^{n} g_{ij}(\gamma(t)) \dot{\theta}^i(t) \dot{\theta}^j(t)} \quad dt$$

A measure class $\mathcal{M}$ is an affine space under its log-likelihood structure, and taking covariances of translations produces a natural Riemannian metric of sorts. Recall that the vector space of translations of $\mathcal{M}$ is a space of random variables R_Ω, and that the tangent space of $\mathcal{M}$ at each point ν is identified with R_Ω, so for each measure ν in $\mathcal{M}$ we can define

$$\langle f, g\rangle_\nu = E_\nu(fg)$$

Clearly this is bilinear, symmetric and positive and would define a Riemannian metric on $\mathcal{M}$ if the expectations existed for every ν and pair of random variables f, g.

A statistical family is a submanifold P of $\mathcal{M}$ with the score $d_p\ell$ identifying the tangent space T_pP at p with a subspace of R_Ω. We therefore define an inner product on T_pP by

$$g_p(v, w) = E_p(d_p(\ell)(v)d_p(\ell)(w))$$

The fact that g is definite follows from the fact that the score is assumed to be injective. This metric is called the *Fisher information* metric. Of course, for the construction to make sense we have to assume that the score takes its values amongst the square-integrable functions.

This construction is an analogue of the natural Riemannian metric on a submanifold of $\mathbf{R}^n$. However it differs from that case in that there the inner product on $\mathbf{R}^n$ at each point is the same whereas in this case the inner product $\langle\ ,\ \rangle_\nu$ at each point of $\mathcal{P}$ changes. We shall give another construction of the Fisher information metric below in section 6.8 which is the direct analogue of the metric on a submanifold in $\mathbf{R}^n$.

For the Riemannian metric on a statistical family P constructed above, the matrix g_{ij} corresponds to the Fisher information. Indeed, if $(\theta^1, \ldots, \theta^n)$ are co-ordinates on P then, since $\partial\ell/\partial\theta^i$ is by definition $d\ell(\partial/\partial\theta^i)$, the components of g in these co-ordinates are

$$g_{ij}(p) = E_p(\frac{\partial\ell}{\partial\theta^i}\frac{\partial\ell}{\partial\theta^j})$$

which is the Fisher information matrix. Note that for the purposes of calculation it is often easier to use the formula

$$g_{ij}(p) = -E_p(\frac{\partial^2\ell}{\partial\theta^i\partial\theta^j})$$

derived from the identity in section 3.2.2.

Example 6.2.1 *Exponential families.* If P is an exponential family then we have

$$\frac{\partial \ell}{\partial \theta^i} = x^i - \frac{\partial K}{\partial \theta^i}$$

Rather than try to calculate the expectation of the product of two of these it is easier to differentiate again and deduce that

$$\begin{aligned} g_{ij}(p) &= - E_p(\frac{\partial^2 \ell}{\partial \theta^i \partial \theta^j}) \\ &= \frac{\partial^2 K}{\partial \theta^i \partial \theta^j} \end{aligned}$$

Example 6.2.2 *The normal family.* Consider the normal family. Choose as co-ordinates μ and σ^2; then we have

$$\frac{\partial \ell}{\partial \mu} = \frac{x - \mu}{\sigma}$$

and

$$\frac{\partial \ell}{\partial \sigma} = \frac{(x-\mu)^2}{\sigma^3} - \frac{1}{\sigma}$$

It is straightforward to calculate the metric now: we obtain

$$g(\frac{\partial}{\partial \mu}, \frac{\partial}{\partial \sigma}) = 0$$

$$g(\frac{\partial}{\partial \mu}, \frac{\partial}{\partial \mu}) = \frac{1}{\sigma^2}$$

and

$$g(\frac{\partial}{\partial \sigma}, \frac{\partial}{\partial \sigma}) = \frac{2}{\sigma^2}$$

This means that the normal family $\mathcal{N}$ can be identified with the upper half plane

$$H = \{(x, y) | y > 0\}$$

with the metric

$$g(\frac{\partial}{\partial x}, \frac{\partial}{\partial y}) = 0$$

$$g(\frac{\partial}{\partial x}, \frac{\partial}{\partial x}) = \frac{1}{y^2}$$

and

$$g(\frac{\partial}{\partial y}, \frac{\partial}{\partial y}) = \frac{2}{y^2}$$

The geometry of this Riemannian manifold is hyperbolic geometry, which was the first non-Euclidean geometry to be discovered Helgason (1978). It is striking that the simplest non-trivial family of distributions should give rise to the simplest non-Euclidean Riemannian manifold.

Example 6.2.3 *The unit normal family.* For the unit normal family we have

$$\ell(\mu)(x) = \frac{1}{2}\sum_i (x - \mu)^2$$

so that the Fisher information metric in this case is just the usual metric on $\mathbf{R}^n$ namely

$$g_{ij} = \delta_{ij}$$

which is the usual Euclidean metric

$$g(v, w) = \sum_i v^i w^i$$

6.3 Metric preserving connections

Consider the question 'Can one choose a connection ∇ such that, under the parallel translation which it defines, the Fisher information is constant, i.e. its value on any pair of tangent vectors remains the same on their parallel translates?' The idea is to make the Fisher information appear as a kind of statistical background which sets the measure of change in statistical properties. We are to think of a random variable f and measure $d\nu$ as a tangent vector at $d\nu$, i.e. as defining a tangent vector to variations through $d\nu$ via the canonical representative $\exp(tf)d\nu$. We want it to be a characteristic property of 1-parameter families of random variables $f(t)$ and $g(t)$, that when they can be regarded as constant vector fields along a 1-parameter family of measures $\nu(t)$, according to some connection ∇, then their covariances $E_{\nu(t)}(f(t)g(t))$ are independent of t. This requires making the proper choice of connection

∇. We must therefore explore how the rate of change of the values of a bilinear function, such as the covariance, is influenced by the rate of change of the two vector fields appearing in its argument, and the choice of connection which defines their rates of change.

Suppose $\gamma : [0,1] \to M$ is a variation in a manifold M starting at $p = \gamma(0)$ and ending at $q = \gamma(1)$. Let ∇ be a given connection on the tangent bundle of M. Given a basis $v_1, v_2, \ldots v_n$ for the tangent space T_pM we can consider its parallel translation along γ which we denote $V_1, V_2, \ldots V_n$. These are vector fields along γ satisfying

$$\nabla V_i(\gamma'(t)) = 0 \qquad V_i(\gamma(0)) = v_i$$

for each $i = 1, 2, \ldots n$.

Let g be a Riemannian metric on M. The property that we require of the connection ∇, is that if X and Y are parallel along γ according to it, then $g(X, Y)$ is constant along γ. In particular, this requires $g(V_i, V_j)$ to be constant for each $i, j = 1, 2, \ldots n$. Given arbitrary vector fields X and Y we can expand them

$$\begin{aligned} X(\gamma(t)) &= f_1(t)V_1(\gamma(t)) + \ldots + f_n(t)V_n(\gamma(t)) \\ Y(\gamma(t)) &= g_1(t)V_1(\gamma(t)) + \ldots + g_n(t)V_n(\gamma(t)) \end{aligned}$$

so that

$$g(X,Y)(\gamma(t)) = \sum_{i,j=1}^{n} g\left(V_i(\gamma(t)), V_j(\gamma(t))\right) f_i(t)g_j(t)$$

Differentiating with respect to t shows

$$\begin{aligned} d\left(g(X,Y)\right)\left(\gamma'(t)\right) = \sum_{i,j=1}^{n} & g\left(V_i(\gamma(t)), V_j(\gamma(t))\right) f_i'(t)g_j(t) \\ & + g\left(V_i(\gamma(t)), V_j(\gamma(t))\right) f_i(t)g_j'(t) \end{aligned}$$

On the other hand, we have

$$\nabla X(\gamma'(t)) = \sum_{i=1}^{n} f_i'(t)V_i(\gamma(t))$$

and

$$\nabla Y(\gamma'(t)) = \sum_{i=1}^{n} g_i'(t)V_i(\gamma(t))$$

so we conclude that

$$d\,(g(X,Y))\,(\gamma'(t)) = g\,(\nabla X(\gamma'(t)), Y(\gamma(t))) + g\,(X(\gamma(t)), \nabla Y\,(\gamma'(t)))$$

If this relationship is to work for all variations γ then it must hold with any tangent vector v in place of $\gamma'(t)$, and we may then depict this relation as an identity between functions depending on three vector fields:

$$d\,(g(X,Y))\,(Z) = g(\nabla X(Z), Y) + g(X, \nabla Y(Z)) \qquad (6.3.1)$$

This condition is necessary if $\nabla X(\gamma'(t)) = 0$ and $\nabla Y(\gamma'(t)) = 0$ are to imply that $g(X,Y)(\gamma(t))$ is constant. However, it is clear that it is also sufficient.

Given a co-ordinate system $\theta^1, \ldots, \theta^n$ we can choose $X = \partial/\partial\theta^i$, $Y = \partial/\partial\theta^j$ and $Z = \partial/\partial\theta^k$. Then $g(\partial/\partial\theta^i, \partial/\partial\theta^j))$ is denoted g_{ij} and

$$dg(X,Y)(Z) = \frac{\partial g_{ij}}{\partial\theta^k}$$

By definition of the Christoffel symbols Γ^k_{ij} we have

$$\nabla \frac{\partial}{\partial\theta^i}(\frac{\partial}{\partial\theta^j}) = \sum_{k=1}^{n} \Gamma^k_{ij} \frac{\partial}{\partial\theta^k}$$

It follows that

$$\begin{aligned}\frac{\partial g_{ij}}{\partial\theta^k} &= g\left(\sum_{l=1}^{n} \Gamma^l_{ik} \frac{\partial}{\partial\theta^l}, \frac{\partial}{\partial\theta^j}\right) + g\left(\frac{\partial}{\partial\theta^i}, \sum_{l=1}^{n} \Gamma^l_{jk} \frac{\partial}{\partial\theta^l}\right) \\ &= \sum_{l=1}^{n} \Gamma^l_{ik} g_{lj} + \Gamma^l_{jk} g_{il}\end{aligned}$$

Permuting the indices i, j, k cyclically we find

$$\begin{aligned}\frac{\partial g_{jk}}{\partial\theta^i} &= \sum_{l=1}^{n} \Gamma^l_{ji} g_{lk} + \Gamma^l_{ki} g_{jl} \\ \frac{\partial g_{ki}}{\partial\theta^j} &= \sum_{l=1}^{n} \Gamma^l_{kj} g_{li} + \Gamma^l_{ij} g_{kl}\end{aligned}$$

If we assume that the connection is symmetric, then Γ^l_{ij} is symmetric in the lower indices ij, and g_{ij} is symmetric also. Subtracting the second equation from the first and adding the third produces

$$\frac{\partial g_{ij}}{\partial \theta^k} - \frac{\partial g_{jk}}{\partial \theta^i} + \frac{\partial g_{ki}}{\partial \theta^j} = 2\sum_{l=0}^{n} \Gamma^l_{ik} g_{jl}$$

By thinking of the subscripts on Γ^l_{ik} as fixed, the right-hand side multiplies the resulting column vector indexed by l by the symmetric matrix g_{lj}, and the whole equation can be regarded as a system of simultaneous equations for the Γ^l_{ik}. These equations have a unique solution, since g_{jl} is an invertible matrix. It is customary to define

$$[ij:k] = 1/2(\frac{\partial g_{ij}}{\partial \theta^k} - \frac{\partial g_{jk}}{\partial \theta^i} + \frac{\partial g_{ki}}{\partial \theta^j})$$

These are called Christoffel symbols of the first kind. The Γ^k_{ij} are called Christoffel symbols of the second kind, and are given by

$$\Gamma^k_{ij} = [ij;l]g^{lk}$$

where g^{lk} is the inverse of the matrix g_{lk}.

It follows that, given any Riemannian metric on a manifold, there is a unique symmetric connection which is metric preserving, in the sense that it preserves the inner products of tangent vectors under parallel translation. It is called the *Levi-Civita* connection. For Euclidean space, if $x^1, \ldots x^n$ is a Cartesian co-ordinate system, then $\partial/\partial x^1, \ldots \partial/\partial x^n$ is an orthonormal system with respect to the standard Riemannian metric, and so the metric tensor g_{ij} is the identity matrix. The partial derivatives of its entries, and hence the Christoffel symbols of the first kind, are all zero, and so therefore are the Γ^k_{ij}. It follows that $\partial/\partial x^1, \ldots \partial/\partial x^n$ are parallel vector fields, and that the metric-preserving connection is indeed the standard Euclidean connection in this case. It is also what we originally called the Levi-Civita connection, so that our two usages of this terminology coincide in this case. Later we will show that the metric preserving connection of a submanifold is obtained from that of its containing Riemannian manifold by orthogonal projection of its rates of change onto the tangent spaces of the submanifold, i.e. by the Levi-Civita connection as defined in Chapter 5. The two views of Levi-Civita connection therefore coincide in general.

Example 6.3.1 *The Levi-Civita connection on the sphere.* As an example, let us determine the Levi-Civita connection of the sphere with its usual Riemannian metric. We show in 6.5.1 below that the trinomial distribution with its Fisher information is the same as the positive octant of the sphere. So the restriction of this connection to the positive octant of the sphere is the 0-connection of the trinomial family. In spherical co-ordinates the bilinear form is given by

$$g(v,w) = \cos^2\phi d\theta(v)d\theta(w) + d\phi(v)d\phi(w)$$

so the matrix g_{ij} is

$$\begin{pmatrix} \cos^2\phi & 0 \\ 0 & 1 \end{pmatrix}$$

Consider the Fisher information metric. We calculate that

$$\begin{aligned} dg(X,Y)(Z) &= Z(E(X(\ell)Y(\ell)) \\ &= E(ZX(\ell)Y(\ell)) + E(X(\ell)ZY(\ell)) + E(X(\ell)Y(\ell)Z(\ell)) \\ &= g(\nabla X(Z),Y) + g(X,\nabla Y(Z)) + E(Z(\ell)Y(\ell)X(\ell)) \end{aligned}$$

where ∇ is the 1-connection. We conclude from this that the 1-connection does not preserve the connection. However, we can easily re-arrange this equation to give

$$\begin{aligned} dg(X,Y)(Z) = g(\nabla X(Z) + \frac{1}{2}E(Z(\ell)Y(\ell)X(\ell))) \\ +g(X,\nabla Y(Z)) + \frac{1}{2}E(Z(\ell)Y(\ell)X(\ell))) \end{aligned}$$

So if we define a connection ∇^0 by

$$\nabla^0 Y(X) = \pi_p(XY(\ell) + \frac{1}{2}X(\ell)Y(\ell))$$

then this is a connection that preserves the metric. Notice that this is, in fact, Amari's 0-connection. It is straightforward to calculate that the Christoffel symbols of this connection are

$$\Gamma^{0n}_{\ ij} = \sum_{mn} g^{nm}(E_p(\frac{\partial^2\ell}{\partial\theta^i\partial\theta^j}\frac{\partial\ell}{\partial\theta^m}) + \frac{1}{2}E_p(\frac{\partial\ell}{\partial\theta^i}\frac{\partial\ell}{\partial\theta^j}\frac{\partial\ell}{\partial\theta^m}))$$

From this we see that it is symmetric and hence the Levi-Civita connection of the Fisher information.

6.4 Dual connections

An inner product g on a vector space allows each vector v to play the role of a linear function, namely $w \mapsto g(v, w)$. It is a perfect correspondence in that every linear function can be uniquely represented in this way. A Riemannian metric allows such a perfect correspondence to be set up between tangent vectors and cotangent vectors on each tangent space. If g is a Riemannian metric on a manifold M, then each vector field X gives rise to a 1-form $\hat{X}$, whose value on any vector field Z is

$$\hat{X}(Z) = g(X, Z)$$

In other words, $\hat{X}$ is just the operation of taking inner products against X. The perfect correspondence between vector fields and 1-forms set up in this way is called a duality, and the vector field X and its associated 1-form $\hat{X}$ are called the duals of each other. It is customary on Euclidean space to use the Euclidean metric to identify 1-forms with vector fields under this duality. For example, the rate of change of a function f is the 1-form df, but it is almost always understood in terms of its dual, the gradient vector field of f. By definition this is the vector field ∇f which realises df by taking inner products against it. It is therefore determined by the relationship

$$df(v) = g(\nabla f(p), v)$$

for each tangent vector v in T_pM.

Recall from section 4.7 that given a connection ∇ on the tangent bundle there is a naturally associated connection on the cotangent bundle which we denote by the same symbol ∇. Given any 1-form ω its rate of change along a vector field Z is another 1-form $\nabla\omega(Z)$. Given a Riemannian metric g and a vector field X, $\nabla\hat{X}(Z)$ is a 1-form, and we can consider its dual, which we denote $\hat{\nabla}X(Z)$, to denote the dependence of this construction on the vector fields X and Z, and to suggest that it defines a connection on vector fields. Because the duality determined by a Riemannian metric it is a perfect correspondence, and $\nabla\hat{X}(Z)$ is linear in Z over smooth functions, $\hat{\nabla}X(Z)$ is linear over smooth functions also. Similarly it is easy to see that $\hat{\nabla}X$ is additive in X. For any smooth function

f and vector field Y we have

$$\begin{aligned} g\left(\hat{\nabla}(fX)(Z), Y\right) &= \nabla(\hat{fX})(Z)(Y) = \nabla(f\hat{X})(Z)(Y) \\ &= df(Z)\hat{X}(Y) + f\nabla\hat{X}(Z)(Y) \\ &= df(Z)g(X,Y) + fg(\hat{\nabla}X(Z), Y) \\ &= g(df(Z)X + \hat{\nabla}X(Z), Y) \end{aligned}$$

It follows that

$$\hat{\nabla}(fX) = dfX + f\hat{\nabla}X$$

and hence $\hat{\nabla}$ is indeed a connection on vector fields. It is called the dual connection of ∇.

The connection ∇ on 1-forms ω is defined in terms of ∇ on vector fields X by the requirement that the rate of change of the function $\omega(Y)$ is given by a Leibniz rule

$$d\left(\omega(Y)\right)(Z) = \nabla\omega(Z)(Y) + \omega\left(\nabla Y(Z)\right)$$

Given the definition of $\hat{\nabla}X$, the formula

$$d\left(\hat{X}(Y)\right)(Z) = \nabla\hat{X}(Z)(Y) + \hat{X}\left(\nabla Y(Z)\right)$$

translates into

$$d\left(g(X,Y)\right)(Z) = g\left(\hat{\nabla}X(Z), Y\right) + g\left(X, \nabla Y(Z)\right) \tag{6.4.1}$$

Formula (6.4.1) proves two things. First it shows, by its symmetry in ∇ and $\hat{\nabla}$, that ∇ is the dual connection of ∇, i.e. that $\hat{\hat{\nabla}} = \nabla$. Secondly it shows that self duality is the same as being metric preserving. If a symmetric connection satisfies $\hat{\nabla} = \nabla$ then it is the Levi-Civita connection for the Riemannian metric which defines the duality. Moreover, among symmetric connections, the self dual connection is unique.

In the previous section we showed that

$$dg(X,Y)(Z) = g(\nabla X(Z), Y) + g(X, \nabla Y(Z)) + E(Z(\ell)Y(\ell)X(\ell))$$

Morover in section 4.6 we defined Amari's α-connection and it is straightforward to use the definition there to show that

$$\nabla^{\alpha}Y(X) = \pi(X(Y(\ell)) + \frac{1-\alpha}{2}X(\ell)Y(\ell))$$

Combining these two results we see that

$$dg(X, Y)(Z) = g(\nabla^{\alpha} X(Z), Y) + g(X, \nabla^{-\alpha} Y(Z))$$

and hence we deduce that the dual to the α-connection is the $(-\alpha)$-connection.

Recall from Chapter 5 that the curvature of a connection ∇ on the tangent bundle is a linear map $T_p M \to T_p M$ for each pair (v, w) of tangent vectors located at p, defined by the equation

$$R(v, w)X = \nabla^2 X(v, w)$$

where the 2 here denotes the fact that the covariant derivative is being taken twice, not that this is the 2-connection. If we write equation (6.4.1) as a relation between 1-forms

$$dg(X, Y) = g(\nabla X, Y) + g(X, \hat{\nabla} Y)$$

then we can apply d again to both sides. Recalling that $d^2 = 0$ and taking care with the fact that 1-forms anti-commute we obtain

$$0 = g(\nabla^2 X, Y) + g(\nabla X, \hat{\nabla} Y) - g(\nabla X, \hat{\nabla} Y) + g(X, \hat{\nabla}^2 Y)$$

and conclude that

$$0 = g(R(v, w)X, Y) + g(X, \hat{R}(v, w)y)$$

It follows that a connection is flat if and only if its dual is flat.

One can prove this in a more computational, co-ordinate based way using formula 6.4.1. Choosing a co-ordinate system $\phi^1, \ldots, \phi^n$ and putting $X = \partial/\partial\phi^i$, $Y = \partial/\partial\phi^j$ and $Z = \partial/\partial\phi^k$ gives

$$\frac{\partial g_{ij}}{\partial \phi^k} = \sum_{l=1}^{n} \hat{\Gamma}^l_{ik} g_{lj} + \Gamma^l_{jk} g_{il}$$

where Γ^k_{ij} and $\hat{\Gamma}^k_{ij}$ are the Christoffel symbols for ∇ and $\hat{\nabla}$ in this co-ordinate system. If ∇ is flat then we may choose $\phi^1, \ldots, \phi^n$ to be affine so that $\Gamma^k_{ij} = 0$ for all indices, and we find

$$\hat{\Gamma}^l_{ik} = \sum_{j=0}^{n} \frac{\partial g_{ij}}{\partial \phi^k} g^{jl}$$

This shows that the co-ordinates $\phi^1, \ldots, \phi^n$ are not affine for $\hat{\nabla}$. However, we can calculate the curvature of $\hat{\nabla}$ from these formulae to show that it is zero, and we leave this as an exercise.

Duality, however, does not preserve symmetry of connections.

6.5 Statistical divergences

Amari has shown that there is a profound geometrical relationship between the Fisher information, considered as a Riemannian metric, the α-family of connections and statistical divergences such as the the relative entropy and Kullback–Leibler distance. In Euclidean space the function $\psi = 1/2((x^1)^2 + (x^2)^2 + \ldots + (x^n)^2)$, where $x^1, \ldots, x^n$ are Cartesian co-ordinates, arises in two different ways from the Riemannian metric

$$g(v, w) = dx^1(v)dx^1(w) + \ldots dx^n(v)dx^n(w),$$

which can be considered to be its infinitesimal counterpart. On the one hand it is clearly one half the length of the geodesic (that is straight line) connecting the origin to a point with co-ordinates $x^1, \ldots, x^n$. On the other hand, its differential is $d\psi = x^1dx^1 + x^2dx^2 + \ldots x^ndx^n$. Under the Euclidean connection ∇, in which $dx^1, \ldots dx^n$ are constant 1-forms, we have $\nabla d\psi = dx^1dx^1 + \ldots + dx^ndx^n$, i.e. $\nabla d\psi$ is twice the Riemannian metric. Amari interprets statistical divergences in this second way, as functions whose second differentials produce the Riemannian metric, in his case the Fisher information.

If θ is a 1-form then $\nabla\theta$ is a 1-form with values which are 1-forms. For example, $d\psi$ is a 1-form whose value on a tangent vector w located at a point p is $x^1(p)dx^1(w) + \ldots + x^n(p)dx^n(w)$. $\nabla d\psi(v)$ is the rate of change of $d\psi$ along the tangent vector v which, if v is located at p, is itself a linear function on T_pP, namely $dx^1(v)dx^1 + \ldots + dx^n(v)dx^n$. Clearly a 1-form whose values are 1-forms is just a bilinear form in which we are evaluating its arguments one at a time. In our example

$$\nabla d\psi(v)(w) = dx^1(v)dx^1(w) + \ldots + dx^n(v)dx^n(w)$$

A Riemannian metric g is a bilinear form, and so it is possible that there may be a 1-form θ such that $\nabla\theta = g$, in the sense that $\nabla\theta(v)(w) = g(v, w)$ for all tangent vectors v and w located at the same point.

In the case of a symmetric connection, if such a 1-form θ can be found, then by the symmetry of g we have

$$d\theta(v, w) = \nabla\theta(v)(w) - \nabla\theta(w)(v) = g(v, w) - g(w, v) = 0$$

Hence, by the Poincáre lemma θ is exact, i.e. has the form $d\psi$ for some function ψ.

The statistical divergence on a statistical family P, determined by a given Riemannian metric g and symmetric connection ∇, is a function $\psi(p,q)$ of pairs of points of P such that, if ψ_p is the function defined by $\psi_p(q) = \psi(p,q)$ (i.e. by holding p fixed) then

$$\begin{aligned} \nabla d\psi_p &= g \\ d\psi_p(v) &= 0 \quad \text{for } v \text{ in } T_pP \\ \psi_p(p) &= 0 \end{aligned}$$

A statistical divergence can be constructed by finding, for each point p, a solution θ_p of the equation $\nabla\theta = 0$ which satisfies $\theta_p(v) = 0$ for every v in T_pP, i.e. θ_p is zero on T_pP. Such a θ is unique, because any two of them differ by a 1-form α satisfying $\nabla\alpha = 0$ and whose restriction to T_pP is zero. By parallel translation, since α is constant, its value on every other tangent space will also be zero. Given θ_p there is a function ψ_p satisfying $d\psi_p = \theta_p$ which is uniquely determined by the requirement that $\psi_p(p) = 0$. Setting $\psi(p,q) = \psi_p(q)$ provides the statistical divergence.

Notice that by its definition ψ_p has a critical point at p with the value zero there. Since its Hessian form $\nabla d\psi_p$ is by definition g, which is positive definite, this critical point is a minimum value. ψ_p can have no other criticial points, and so it is a convex function with a unique minimum value of 0 occurring at p. Thus $\psi(p,q) \geq 0$ for every p and q, with $\psi(p,q) = 0$ if and only if $p = q$.

The condition that $\nabla\theta = g$ should have, for each point p, a solution which is zero on T_pP, forces the connection to be flat. Recall from Chapter 5 that an n-dimensional family of solutions of the equation $\nabla\alpha = 0$, where n is the dimension of P, provides an affine co-ordinate system for ∇. We can generate such a family by taking the differences between the solutions of the equation $\nabla\theta = g$, or rather the infinitesimal differences around a point p. For a given tangent vector v in T_pP let $\gamma(t)$ be the corresponding geodesic of ∇ through p, and consider the 1-parameter family of 1-forms on P, namely $\theta_{\gamma(t)}$ We denote the derivative at $t = 0$ of this family by α_v, i.e

$$\alpha_v(w) = \lim_{t \to 0} \frac{\theta_{\gamma(t)}(w) - \theta_p(w)}{t}$$

For each v the 1-form α_v satisfies $\nabla\alpha_v = 0$, and it depends linearly on v. It is easy to show that $\alpha_v \neq 0$ for any $v \neq 0$, and so the α_v's are an n-dimensional family.

While flatness of ∇ is a necessary condition for the existence of a statistical divergence, it is not sufficient. Flatness means that $\nabla^2\theta = 0$ for any 1-form θ, and so if $\nabla\theta = g$ is to have a solution then necessarily

$$\nabla g = \nabla^2\theta = 0$$

Since ∇ is flat we may chose an affine co-ordinate system for it $\phi^1, \ldots, \phi^n$, i.e. one for which $\nabla d\phi^i = 0, i = 1, \ldots n$. We may express g as

$$\sum_{i,j=1}^{n} g_{ij}\, d\phi^i d\phi^j = \sum_{j=1}^{n} \left(\sum_{i=1}^{n} g_{ij}\, d\phi^i \right) d\phi^j$$

where the last expression treats g as a 1-form with values in 1-forms. As a 1-form it has an expansion $a_1 d\phi^1 + \ldots + a_n d\phi^n$, and since its values are 1-forms each coefficient a_j is itself a 1-form. Moreover

$$\nabla(a_j d\phi^j) = \nabla(a_j) \wedge d\phi^j + a_j d^2\phi^j$$

and since the second term in the sum is zero we have

$$\nabla g = \sum_{j=1}^{n} \nabla \left(\sum_{i=1}^{n} g_{ij}\, d\phi^i \right) \wedge d\phi^j$$

Using the fact that $d\phi^i$ is covariantly constant for each i, we have

$$\nabla \left(\sum_{i=1}^{n} g_{ij}\, d\phi^i \right) = \sum_{i=1}^{n} dg_{ij}\, d\phi^i$$

and so

$$\nabla g = \sum_{i=1}^{n} \left(\sum_{j,k=1}^{n} \frac{\partial g_{ij}}{\partial \phi^k} d\phi^k \wedge d\phi^j \right) d\phi^i$$

It follows that the condition for $\nabla g = 0$ is

$$\frac{\partial g_{ij}}{\partial \phi^k} = \frac{\partial g_{ik}}{\partial \phi^j}$$

This is precisely the condition for the dual connection $\hat{\nabla}$ to be symmetric.

By following a Poincaré lemma style of argument it follows that the symmetry of $\hat{\nabla}$ is also sufficient for solutions of the equation $\nabla\theta = g$ to exist, and with arbitrarily prescribed initial conditions. Hence we arrive at the result, that a statistical divergence determined by a Riemannian metric g and connection ∇ exists if and only if ∇ (and hence $\hat{\nabla}$) are flat and both ∇ and $\hat{\nabla}$ are symmetric. Amari (1985) proved this result assuming that ∇ is flat. We have further shown here the necessity of ∇ being flat.

6.5.1 *Orthogonality and divergence minimisation*

Given a distribution p in a statistical family one often wants to choose a distribution from some subfamily S which in a sense best approximates it, or is a best estimate. In statistics the idea of best is often interpreted as minimum divergence. Divergences are expected to behave in a sufficiently geometric way that the point on the subfamily S of minimum divergence from p lies on some kind of trajectory from p to S which is orthogonal to S. The paradigm for this kind of thinking is a point on a surface of minimum distance from a given point p not on the surface. Such a point is always at the foot of a perpendicular to the surface from p. Amari (1085) shows how, in the framework of his theory of duality and divergences, this geometric interpretation works using the statistical divergence in place of distance, if one uses the geodesics of the dual connection $\hat{\nabla}$ as the substitute for Euclidean straight lines.

In order that a point q in a submanifold S be the location of a minimum value of a given function f over S, df must be zero on the tangent space to S at q. (We will not consider the case where a minimum value occurs on the boundary of a region.) For if v is a tangent vector to S at q then it is realised by a variation γ through q which lies entirely in S. Since $\gamma(0)$ is the location of the minimum value of f it follows that $f(\gamma(t))$ has a minimum at $t = 0$, and that $df(v) = (f \circ \gamma)'(0) = 0$. This condition can always be interpreted geometrically by considering the gradient vector field ∇f of f. By definition it is the vector field which realises df by taking inner products of tangent vectors against it, i.e. such that

$$df(v) = g(\nabla f, v)$$

The condition that df should vanish on T_qS translates into the condition that $\nabla f(q)$ be orthogonal to T_qS.

One can describe this condition in terms of curves being orthogonal to T_qS by considering the integral variations of the gradient vector field of f. The integral variations of any vector field X are variations γ whose tangent vector at any point is given by the value of the vector field there. In other words, γ must satisfy

$$\gamma'(t) = X(\gamma(t))$$

for every t. Applying $d\phi^1, \ldots d\phi^n$ to this equation, where $\phi^1, \ldots, \phi^n$ is a co-ordinate system, we have

$$\frac{d}{dt}\phi \circ \gamma(t) = d\phi^i(\gamma'(t)) = d\phi^i\left(X(\gamma(t))\right) \quad i = 1, \ldots n$$

Remember that $\phi^i \circ \gamma(t)$ is usually abbreviated $\phi^i(t)$ and that $d\phi^i(X)$ is the component function X^i in the expansion $X = X^1(\partial/\partial\phi^1) + \ldots + X^n(\partial/\partial\phi^n)$. Hence the defining relation for an integral variation takes the co-ordinate form

$$\dot{\phi}^i = \bar{X}^i(\phi^1, \ldots \phi^n) \quad i = 1, \ldots, n$$

where $\bar{X}^i$ is the co-ordinate form of the function X^i. This is a system of ordinary differential equations, and the standard existence and uniqueness theorems show that solutions exist over suitably small time intervals and are unique. For the sake of simplicity we will assume that solutions exist for all time. What the existence and uniqueness theory says is that, for any smooth vector field X, through every point of the manifold there passes a unique integral variation of X.

Given a function f, the integral variations of its gradient vector field trace out a family of curves in a direction of increasing values of f. Indeed, if γ is an integral variation of ∇f then

$$(f \circ \gamma)'(t) = df(\gamma'(t)) = g\left(\nabla f(\gamma(t)), \nabla f(\gamma(t))\right) \geq 0$$

The curves traced out by the integral variations of the gradient vector field are called the paths of steepest descent of f, being oriented in the direction of decreasing value. Since tangents to these curves are in the direction of ∇f at each point, in order that q be a minimum value of f over a submanifold S the curve of steepest descent through q must be orthogonal to S.

Suppose that, like a statistical divergence, the minimum value of the function f over the entire manifold P occurs at a unique point p. Along its curves of steepest descent the values of the function are strictly decreasing. The values must therefore approach the minimum value, and be confined to a compact region of P containing the unique location p where this minimum value occurs. From here it is easy to show that p must be a limit of the curves of steepest descent, so that in effect all curves of steepest descent radiate out from p. Hence we can say that for a statistical divergence $\psi(p,q)$, the curves of steepest descent of ψ_p all radiate in to p, and if q is the point on any submanifold S at which $\psi(p,q)$ achieves its minimum value, then the curve of steepest descent frorm p through q must cut orthogonally through S.

For Amari's theory it remains to identify the curves of steepest descent for ψ_p with the geodesics of $\hat{\nabla}$ emanating from p. Let us put $\theta = d\psi_p$ and $X_\theta = \nabla\psi_p$ so that X_θ is the dual of θ and the gradient vector field of ψ_p. Its integral variations trace out the curves of steepest descent for ψ_p. Considering the definition of $\hat{\nabla}$, we have

$$\nabla\theta(Y)(Z) = g\left(\hat{\nabla}X_\theta(Y), Z\right)$$

for any two vector fields Y and Z. Hence for any vector field Y, since θ is a solution of the equation $\nabla\theta = g$, we have

$$\hat{\nabla}X_\theta(Y) = Y$$

If we compute the rate of change under $\hat{\nabla}$ of some multiple fX_θ along X_θ itself, then since $\hat{\nabla}(X_\theta) = X_\theta$ we find

$$\hat{\nabla}(fX_\theta)(X_\theta) = df(X_\theta)X_\theta + fX_\theta$$

The equation $df(X_\theta) + f = 0$ is a first-order partial differential equation which can always be solved in these circumstances, indeed via the method of characteristics which turns it into a family of ordinary differential equations along the integral variations of X_θ. With this choice of f we conclude that

$$\hat{\nabla}(fX_\theta)(fX_\theta) = f(\hat{\nabla}(fX_\theta)(X_\theta)) = f(df(X_\theta) + f)X_\theta = 0$$

If γ is any integral variation of fX_θ then by definition $\hat{\nabla}\gamma'(\gamma') = 0$. Hence the integral variations of fX_θ, i.e. of the rescaled gradient vector field of ψ_p, are geodesics of $\hat{\nabla}$.

It remains only to observe that the integral variations of any multiple fX of a vector field X trace out the same curves as the integral variations of X itself. Indeed, for any variation $\gamma(t)$ the chain rule shows that $\gamma(F(t))$ has tangent vector $F'(t)\gamma'(F(t))$, and it traces out the same curve as $\gamma(t)$. If γ is an integral variation of X then $\gamma(F(t))$ will be an integral variation of fX provided we choose F to satisfy

$$F'(t)\gamma'(F(t)) = f\left(\gamma(F(t))\right) X\left(\gamma(F(t))\right) = f\left(\gamma(F(t))\right)\gamma'(F(t))$$

That is, we must choose F to satisfy the ordinary differential equation

$$F'(t) = f\left(\gamma(F(t))\right)$$

Since this is always possible, the integral variations of fX always have the form $\gamma(F(t))$ where $\gamma(t)$ is an integral variation of X, and so the integral variations of both vector fields trace out the same curves.

In particular, the integral variations of the gradient vector field of a statistical divergence trace out geodesics of $\hat{\nabla}$, since there is a rescaling of it whose integral variations are geodesics. We therefore arrive at Amari's result, that the point q on a submanifold S of a statistical family P, which has minimum divergence from a point p, must have the $\hat{\nabla}$ geodesic from p to q cutting S orthogonally.

6.6 Transformation models

Consider the case when P is a transformation model for a group G. Then it is eay to check that the the log-likelihood commutes with the action of G. That is, if g is an element of G and $T_p(L_g)$ the tangent map at p to the map $L_g(p) = gp$, i.e. the left action of g, so that

$$T_p(R_g)\colon T_pP \to T_{pg}P$$

we have

$$d_{gp}\ell \circ T_p(L_g) = g^{-1}d_p\ell$$

where on the right hand side of this formula the g^{-1} is acting on the space of random variables by $(g^{-1}f)(x) = f(gx)$. Moreover, we have for all random variables f that

$$E_{gp}(f) = E_p(g^{-1}f)$$

Combining these two results shows that for any v and w in T_pP we must have

$$g_p(v, w) = g_{gp}(T_p(L_g)(v), T_p(L_g)(w))$$

or that the Fisher information metric is invariant under the action of G.

We can use this fact to show that the Fisher information metric on the normal family is indeed the hyperbolic metric. The group $SL(2, \mathbf{R})$ of all real unit determinant matrices acts on the complex plane by fractional linear transformations

$$\begin{pmatrix} a & b \\ c & d \end{pmatrix} z = \frac{az + b}{cz + d}$$

Notice that if c and d where complex this expression would not be defined for $z = -d/c$, but if c and d are real and z is in the upper half space so that its imaginary part is strictly positive then it is not possible to have $z = -d/c$. Moreover, because the matrix entries are real this action maps points in the upper half space to points in the upper half space. It therefore acts on the set of all normal distributions when they are parametrised by μ and σ. As we saw in Example 3.4.1, if we restrict to the subgroup of matrices of the form

$$\begin{pmatrix} a & b \\ 0 & 1/a \end{pmatrix}$$

this is the action of the location scale group.

The location scale group also acts on the real line and the action on the normal densities is that induced by the action on the real line. It follows that the Fisher information metric is invariant under the location scale group. The group $SL(2, \mathbf{R})$ acts on the real line if we adjoin a point at infinity. However the induced action on measures does not preserve the normal densities. For instance an inversion

$$x \mapsto \frac{1}{x}$$

does not leave a normal density looking like a normal density! However, it is still true that with the action of $SL(2, \mathbf{R})$ that we have described on $\mathcal{N}$ the Fisher information metric is invariant. The only invariant metric on $\mathcal{N}$ of this type is the hyperbolic metric. To see that there is only essentially one invariant metric consider the tangent space at $N(0,1)$. Then whatever the metric is here determines it everywhere by moving it around with the group. But more than this is true. The metric is also invariant under the action of the subgroup stabilising the point $p = N(0,1)$ which is subgroup $S0(2)$ of rotations. This group acts on the tangent space at p by the maps $T_P(L_g)$ for $p \in SO(2)$. The tangent space is two-dimensional and it is straightforward to check that this action is the standard action of rotations on $\mathbf{R}^2$. It follows that up to scale there is only one inner product on $T_p\mathcal{N}$ invariant under $SO(2)$ and hence, up to scale, only one metric on $\mathcal{N}$ invariant under $SL(2, \mathbf{R})$. This invariance property of the hyperbolic metric is important in differential geometry, but it is not clear if it has any significance in statistics.

6.7 Other geometries

The geometry we have been considering up to now is called the *expected geometry* of a family, relying as it does so much on the taking of expectations. There are two other approaches to the geometry of a statistical manifold which we wish to mention. We do not have sufficient space to deal with them in detail.

Observed geometry was introduced by Barndorff-Nielsen, see for instance Barndorff-Nielsen (1986a). We shall see below in section 6.7.2 that when formulated in terms of *yokes* it, in fact, includes the expected geometry.

The second geometry that we will consider very briefly is the preferred-point geometry of Critchley and Marriott, (Marriott (1991).

6.7.1 *The observed Fisher information metric*

To understand the observed Fisher information metric let us first prove a simple result about functions on a manifold. Let f be a function on P with a local maximum at p. So $d_p f = 0$.

Then we can define a symmetric, bilinear function on the tangent space at p called the Hessian of f. The Hessian is essentially the matrix of second derivatives. We define it by choosing co-ordinates $(\theta^1, \ldots, \theta^r)$ and letting

$$\mathrm{Hess}_p(f) = \frac{\partial^2 f}{\partial \theta^i \partial \theta^j} d\theta^i d\theta^j$$

By $d\theta^i d\theta^j$ we mean the bilinear function on T_pP whose value on the pair of tangent vectors X and Y is

$$d\theta^i d\theta^j(X, Y) = d\theta^i(X) d\theta^j(Y)$$

At first sight this definition goes against all our principles as we had to choose co-ordinates to define it. The point is that because $df_p = 0$ the definition is independent of co-ordinates. Indeed if $(\phi^1, \ldots, \phi^r)$ are another set of co-ordinates on P then we have

$$\begin{aligned} \frac{\partial^2 f}{\partial \theta^i \partial \theta^j}(p) &= \frac{\partial^2 f}{\partial \phi^k \partial \phi^l}(p) \frac{\partial \phi^k}{\partial \theta^i}(p) \frac{\partial \phi^l}{\partial \theta^j}(p) + \frac{\partial f}{\partial \phi^k}(p) \frac{\partial^2 \phi^k}{\partial \theta^i \partial \theta^j}(p) \\ &= \frac{\partial^2 f}{\partial \phi^k \partial \phi^k}(p) \frac{\partial \phi^k}{\partial \theta^i}(p) \frac{\partial \phi^l}{\partial \theta^j}(p) \end{aligned}$$

Using the transformation rule for 1-forms

$$d\theta^i(p) = \frac{\partial \theta^i}{\partial \phi^j}(p) d\phi^j(p)$$

we see that

$$\frac{\partial^2 f}{\partial \phi^k \partial \phi^l}(p) d_p\phi^k d_p\phi^l = \frac{\partial^2 f}{\partial \theta^k \partial \theta^l}(p) d_p\theta^k d_p\theta^l$$

so that $\mathrm{Hess}_p(f)$ is independent of the choice of co-ordinates. Just as in calculus in many variables if p is a maximum for f then $\mathrm{Hess}_p(f)$ is negative definite.

The maximum likelihood estimator (mle)

$$M : \Omega \to P$$

is defined to be that function which assigns to a point x in the sample space the probability density $M(x)$ which is a maximum at x, if that is uniquely defined. We will assume for this section that the mle is a bijection. This may not be the case and we will discuss what happens then in the next chapter where we consider the mle in more detail. We should point out now though that the mle is one point at which we have to depart from standard statistical notation. The statistical notation for the mle is parameter dependent because if $(\theta^1, \ldots, \theta^n)$ are co-ordinates on P then the mle of some point x has co-ordinates $(\hat{\theta}^1(x), \ldots, \hat{\theta}^n(x)$ where the x is usually suppressed. In other words the function $\hat{\theta}$ is the composition of θ and M.

Define a function $Y: P \times P \to \mathbf{R}$ by

$$Y: P \times P \to \mathbf{R}$$

by $Y(p, q) = \ell(p)(M^{-1}(q))$. Notice that the maximum of the function $Y(\ ,p)$ occurs at the point p. Hence we have

$$d_p Y(\ ,p) = 0$$

and the Hessian $H_p(Y(,p))$ is negative definite. The *observed Fisher information metric* is defined to be

$$g_p = -\mathrm{Hess}_p(Y(\ ,p))$$

It is interesting to note that we can also define the expected Fisher information in a similar way. We define first

$$Y(p, q) = E_q(\ell(p))$$

then $d_p Y(,p) = E_p(d\ell_p) = 0$. From this we calulate that in co-ordinates

$$\mathrm{Hess}_p(Y(\ ,p)) = E_p(\frac{\partial^2 \ell}{\partial \theta^i \partial \theta^j}(p) d_p\theta^i d_p\theta^j$$

which is the Fisher information metric.

6.7.2 *Yokes*

The ideas involved in observed geometry have been abstracted by Barndorff-Nielsen and Blæsild in their definition of a *yoke* first introduced in Barndorff-Nielsen (1987). The precise definition of a yoke is that it is a function $Y: P \times P \to \mathbf{R}$ such that

1. $d_p Y(,p) = 0$, and
2. $\mathrm{Hess}_p(Y(,p)$ is negative definite.

The two examples we have considered so far are the expected yoke:

$$Y(p,q) = E_q(\ell(p)),$$

and the observed yoke

$$Y(p,q) = \ell(p)(M^{-1}(q)$$

Any yoke defines a metric by defining the inner product on the tangent space at p to be

$$\mathrm{Hess}_p(Y(\ ,p))$$

We leave it as an exercise for the reader to show that for the observed (expected) yokes this metric is the observed (expected) Fisher information metric.

There is a similar generalisation of the α connections. Consider, as in Exercise 4.1, a function $\phi: P \times P \to TP$ with the property that $\phi(p,q) \in T_p P$. If Y is a yoke then either of $d_p Y(\ ,q)$ or $d_p Y(q,\ )$ has this property. Denote by $\phi_p: P \to T_p P$ the function $\phi_p(q) = \phi(p,q)$. We can now define a connection using ϕ. If Y is a vector field on P then $Y(\phi)$ is a function from P to $T_p P$ and if X is a vector at p then $X(Y(\phi))$ is an element of $T_p P$. If we define

$$\nabla(Y)(X) = X(Y(\phi))$$

it is straightforward to check that this defines a connection. Define

$$\phi_p^\alpha(q) = \frac{(1-\alpha)}{2} d_p Y(\ ,q) + \frac{(1+\alpha)}{2} d_p Y(q,\ )$$

The the α-connection define by the yoke Y is the connection defined by ϕ^α. We leave it as an exercise for the reader to verify that if the yoke is $Y(p,q) = E_q(\ell(p))$ then this definition recovers the expected α-connections. If the yoke is the observed yoke this defines the observed α-connections.

6.7.3 *Preferred point geometry*

Marriott (1987) defined the notion of *preferred point* geometry which was further developed in Critchley, Marriott and Salmon (1991). We will give here only the definition and refer the reader to these references for the interesting results that are obtained. A preferred point geometry is a manifold with a family of metrics g_p. Here g_p is a metric on an open neighbourhood of the point p in M.

6.8 The square root likelihood

When we are given a manifold with a metric specified as a function as a pairing on the tangent spaces is it often difficult to visualise what the 'shape' of the manifold is. For instance which points are close together ? Classically manifolds first arose as surfaces in $\mathbf{R}^3$. For such a surface Σ there is a natural way to define a Riemannian metric. If X and Y are tangent vectors at p in the surface then we just define $g_p(X, Y)$ to be the inner product of X and Y in $\mathbf{R}^3$. This Riemannian structure is easy to visualise as the length of any curve in Σ is just its length in $\mathbf{R}^3$. So the 'shape' of Σ is its shape in $\mathbf{R}^3$. The way we have defined the Fisher information metric we have regarded P as a submanifold in $\mathcal{M}$ which is an affine space so analogous to Euclidean space. As we have already noted, the inner product that we use on each tangent space changes as we move about $\mathcal{M}$ so we do not have the simple picture above. The purpose of this section is to describe an alternative way of defining the Fisher information metric which does enable us to visualise the shape of P.

We introduce the square root of the likelihood function instead of its log and think of it as a map

$$\sqrt{p}: P \to R_\Omega$$

which sends p to the random variable $x \mapsto 2\sqrt{p(x)}$. Notice that this is just the α-embedding of Amari, discussed in section 4.6, in the case that $\alpha = 0$. We shall call this map the square root likelihood. The derivative of the square root likelihood defines a map

$$d\sqrt{p}: T_p P \to R_\Omega$$

It is elementary calculus to show that

$$d\sqrt{p} = \sqrt{p}d\ell. \tag{6.8.1}$$

We can define an inner product on R_Ω by

$$\langle f, g\rangle = \int_\Omega fg\mu$$

for any two random variables f and g. This is, of course the inner product defined using Amari's 0-expectation. Notice that there is no dependence on μ in this definition. Using the map $\sqrt{p}: T_pP \to R_\Omega$ defines an innner product g'_p on T_pP by

$$g'_p(v, w) = \langle d\sqrt{p}(v), d\sqrt{p}(w)\rangle$$

If we calculate we discover that

$$\begin{aligned} g'_p(v, w) =& \langle d\sqrt{p}(v), d\sqrt{p}(w)\rangle \\ =& \int_\Omega d\ell(v)d\ell(w)\mu \\ =& E_p(d\ell(v)d\ell(w)) \end{aligned}$$

This is the Fisher information metric. We shall discuss below how to define $\sqrt{p}$ without reference to μ.

Example 6.8.1 *The trinomial family.* Let us illustrate these ideas with the trinomial family. The sample space Ω is the set of points $\{1, 2, 3\}$ and we take as base measure μ just counting measure; that is the μ-measure of a set is just the number of points in it.

The space R_Ω is identified with the space $\mathbf{R}^3$. Indeed a random variable is just a map f from $\{1, 2, 3\}$ to $\mathbf{R}$ so it is determined by the list of three numbers which are its values $(f(1), f(2), f(3)) = (f^1, f^2, f^3)$. These define a point $(f(1), f(2), f(3)) = (f^1, f^2, f^3)$ in $\mathbf{R}^3$.

The set of all probability distributions P on Ω can then be regarded as a subset of the set of all functions on Ω by comparison with this standard measure. In this way we identify P with the set of functions $p = (p^1, p^2, p^3)$ with $p^1 + p^2 + p^3 = 1$ and each p^i positive. In other words if δ_i is the measure which is a point mass at i then the probability measure corresponding to p is

$$(p_1\delta_1 + p_2\delta_2 + p_3\delta_3)$$

The inner product used in the square root likelihood construction of the Fisher information metric is just four times the standard inner product on $\mathbf{R}^3$. That is

$$\begin{aligned}\langle f, g\rangle &= \int_\Omega fg\mu \\ &= \sum_{i=1}^{3} f^i g^i\end{aligned}$$

The square root likelihood embeds the space P into $\mathbf{R}^3$ as the intersection of the sphere of radius 1 with the positive octant. It follows that the Fisher information metric on P is just four times the metric it gets as this subset of the sphere of radius 1. We can remove the factor of 4 by just doubling the size of the sphere if we wish.

Notice that if we identify $\mathcal{M}$ with $\mathbf{R}^3$ as we did for the binomial family in Example 1.4.1 then $\mathcal{P}$ is the set of all triples (x^1, x^2, x^3) such that

$$\exp(x^1) + \exp(x^2) + \exp(x^3) = 1$$

This has an entirely different shape to the intersection of the sphere with the positive octant. For instance it is an unbounded surface whereas the intersection of the sphere with the positive octant is bounded.

6.8.1 *The Levi-Civita connection and the square root likelihood*

In section 4.4 we showed how to define connections on submanifolds by projection. If P is any Riemannian manifold and Q is a submanifold of P then the restriction of the inner product on T_qP to T_qQ, at any q in Q, defines a Riemmanian metric on Q. This metric can be used to define the Levi-Civita connection on Q. On the other hand by using orthogonal projection we can define a connection induced by the Levi-Civita connection on P. This, in fact, coincides with the Levi-Civita connection on Q. This follows from a simple calculation and the characterisation of the Levi-Civita connection as the unique symmetric connection compatible with the metric.

We have already seen this result in our statistical setting. That is where, the large Riemannian manifold is the space of random variables R_Ω with the inner product $\langle f, g \rangle = \int_\Omega f g d\mu$, and the submanifold is the image of our family under the square root likelihood. Then the induced metric is the Fisher information metric and in Chapter 4 we saw that the connection induced by orthogonal projection was the 0-connection, that is the Levi-Civita connection of the Fisher information metric.

6.9 Integration and densities on a manifold

We want to consider the question of how to integrate on a manifold. Of course, in one sense, this question is answered by measure theory. You just choose a measure for the σ-algebra generated by the domains of co-ordinate charts. However, the point we wish to make is that there is a natural class of measures called densities that are generalisations to manifolds of Lebesgue measure on $\mathbf{R}^n$.

Consider how we could try to define a Riemann integral of a function on a manifold. We would divide the area to be integrated over into polyhedra, evaluate the function at some point in each polyhedron, divide by the area of the polyedron and sum. Then we would take a limit over smaller and smaller divisions. For a general manifold the main problem with this is deciding what the sizes of the small polyhedra are.

In the plane for instance we can consider the same problem. Then we usually decide that the square of side length 1 has area 1. If we take a general parallelogram with sides given by vectors v^1 and v^2 then its area is $|\det(v^i_j)|$ where $v^i = (v^i_1, v^i_2)$. Notice that the general parallelogram is obtained from the square by applying the matrix v^i_j. If we are in a two-dimensional vector space where we don't know what the size of a square is, or indeed even what a square is, then we can just assign to any parallelogram S a positive number $\mu(S)$ subject to the fact that if X is a linear transformation of the space then $\mu(XS)$ equals $|\det(X)|\mu(S)$. A parallelogram is determined by the side vectors v_1 and v_2 and we may as well exclude from this discussion the case when v^1 and v^2 do not form a basis as then the figure has zero volume. We usually define μ as a function of the basis (v_1, v_2) instead of S.

For a general vector space V we now define a *density* to be a map μ from the set of all bases for V into $\mathbf{R}$, such that if X is an invertible matrix then

$$\mu(\sum_{j_1} X^1_{j_1} v^{j_1}, \ldots, \sum_{j_n} X^n_{j_n} v^{j_n}) = |\det(X)|\mu(v^1, \ldots, v^n)$$

Denote by $\Delta(V)$ the set of all densities and note that it is a real, one-dimensional vector space. Call a density positive if it always takes positive values. We shall see that positive densities play the role of measures and densities the role of measures multiplied by functions.

Consider the problem again in $\mathbf{R}^n$ and assume that we have a density chosen for each tangent space to $\mathbf{R}^n$. This then is some μ_x for each x in $\mathbf{R}^n$. We define its integral to be

$$\int_{\mathbf{R}^n} \mu = \int_{\mathbf{R}^n} \mu_x(\frac{\partial}{\partial x^1}, \ldots, \frac{\partial}{\partial x^n}) dx^1 \ldots dx^n$$

We now can identify the usual Lebesgue measure $dx^1 \ldots dx^n$ with the density that sends a basis $v^1, \ldots, v^n$ to $|\det(v^i_j)|$. This definition has an important property as follows. Notice that if $\phi : \mathbf{R}^n \to \mathbf{R}^n$ is a diffeomorphism, i.e. a change of co-ordinates, we can define

$$\phi^*(\mu)_x(v_1, \ldots, v_n) = \mu_{\phi(x)}(T_x(\phi)(v_1), \ldots, T_x(\phi)(v_n))$$

This is a definition that obviously extends to manifolds. With this definition the important property of densities is

$$\int_{\mathbf{R}^n} \phi^* \mu = \int_{\mathbf{R}^n} \mu$$

Proving this is just a question of undoing the definition of the integral of a density and seeing that the usual Jacobian factor in a change of variables cancels out with the factor coming from the transformation of μ.

It is easy to pass now to a general manifold. Notice that

$$\Delta(P) = \cup_{p \in P} \Delta(T_p P)$$

is, like the tangent bundle, a family of vector spaces, one for each point of the manifold. However, in this case, each vector space is one-dimensional. We can make $\Delta(P)$ into a manifold in a similar way to the case of the tangent bundle. We call a density on P a smooth map $\rho: P \to \Delta(P)$ with $\rho(p) \in \Delta(T_pP)$. In other words it is a smooth choice of density for each tangent space. Densities are the things that we can integrate on a manifold.

To integrate on a manifold first consider a density μ with support in a co-ordinate neighbourhood, that is μ is zero outside a co-ordinate neigbourhood. Then the co-ordinates define a diffeomorphism to $\mathbf{R}^n$ so we can pull the density back to a density on $\mathbf{R}^n$ and integrate it using the definition above. If we change the choice of co-ordinates that is the same as applying a diffeomorphism to $\mathbf{R}^n$ and we know that this leaves the integral of a density unchanged. So this definition is independent of the choice of co-ordinates. To integrate a general density we need write it as a sum of densities supported in co-ordinate neighbourhoods and then sum all the corresponding integrals. This we can do if we have some topological assumption such as paracompactness. We also have to check that this definition is independent of the decomposition. All these technical things can be carried through, see for instance Bott and Tu (1982).

In the case of a Riemannian manifold there is a naturally defined density called the Riemannian density or sometimes the Riemannian volume. Because the manifold has a metric there is a distinguished family of bases; those that are orthonormal. If two orthonormal bases are related by a matrix X then it is easy to deduce that

$$XX^t = 1$$

and therefore that $|\det(X)| = 1$. We can therefore define a density on P by defining it to be 1 on any orthonormal basis and letting its value on any other basis be determined by the transformation rule. More geometrically we are declaring that a box with orthogonal edges of length 1 must have volume 1 and because we have the metric we know what it means for the edges to be orthogonal of length 1.

If we take a co-ordinate basis then it is the transform of some orthonormal basis e_i:

$$\frac{\partial}{\partial \theta^i} = \sum_j X_i^j e_j$$

and from this it easily follows that $g_{ij} = (XX^t)_{ij}$ and therefore that $|\det(X)| = \sqrt{|\det(g_{ij}|}$. Hence we have

$$\mu_p(\frac{\partial}{\partial\theta^1}, \ldots, \frac{\partial}{\partial\theta^n}) = \sqrt{|\det(g_{ij}|}$$

or that

$$\mu_p = \sqrt{|\det(g_{ij}|} d\theta^1 \ldots d\theta^n$$

Note that this means that on $\mathbf{R}^n$ with the metric given by a constant, positive definite matrix Σ the Riemannian density is

$$\mu_p = \sqrt{|\det(\Sigma_{ij}|} d\theta^1 \ldots d\theta^n$$

Lastly let us reasure the reader who has read or been told that the things to integrate on a manifold are differential n-forms where n is the dimension of the manifold. What is precisely true is that if you have an oriented manifold then you can integrate differential n forms. However, choosing an orientation amounts to identifying n forms and densities so there is no contradiction with the discussion here.

6.9.1 *Half-densities and the square root likelihood*

To define the square root likelihood without choosing an origin for the space of measures we need to introduce half-densities. These are the answer to the question of what kind of object on a manifold has a square that can be integrated. A half-density on a vector space V is a map ρ from the set of all bases for V into $\mathbf{R}$ such that if X is an invertible matrix then

$$\rho(\sum_{j_1} X^1_{j_1} v^{j_1}, \ldots, \sum_{j_n} X^n_{j_n} v^{j_n}) = |\det(X)|^{1/2} \rho(v^1, \ldots, v^n)$$

Clearly if ρ and η are two half densities then their product is a density. We denote by $\Delta^{1/2}(V)$ the vector space of all half-densities.

For a manifold a half-density is a smooth choice of half-density on each tangent space.

Consider now how this applies in our statistical setting. If the sample space Ω is a manifold then denote by $\mathcal{M}^{1/2}$ the space of all half-densities. Notice that this vector space has a natural inner product defined by

$$\langle \rho, \eta \rangle = 4 \int_M \rho\eta$$

We say that a statistical manifold admits a square root if there is a map

$$P \to \mathcal{M}^{1/2}$$

usually denoted $p \mapsto \sqrt{p}$ such that the product $\sqrt{p}\sqrt{p}$ is the measure p. We shall generally assume that this map is in fact a map from all of $\mathcal{M}$ to $\mathcal{M}^{1/2}$. We shall call this map the square root likelihood. In such a case the image of P under the square root likelihood inherits a metric as a submanifold of $\mathcal{M}^{1/2}$. To calculate what this is consider the tangent map to the square root likelihood. If f is a tangent vector to P at p then it is tangent to the curve $p + tf = e^{tf}p$ and this is mapped to $e^{1/2tf}\sqrt{p} = \sqrt{p} + 1/2tf\sqrt{p} + O(t^2)$. Hence the tangent vector f is mapped to $1/2f\sqrt{p}$. So the inner product of a pair of tangent vectors f and g is given by

$$4\int_\Omega \frac{1}{2}f\sqrt{p}\frac{1}{2}g\sqrt{p} = \int_\Omega fgp$$

or the Fisher information metric.

6.10 Remarks for Chapter 6

Remark 6.1 A good reference for Riemannian geometry is Helgason (1978).

6.11 Exercises for Chapter 6

Exercise 6.1 Show that the curves traced out by geodesics in the normal family $\mathcal{N}$, thought of as the upper half-space

$$\mathcal{N} = \{(\mu, \sigma) \mid \sigma > 0)\}$$

are either vertical lines or half-circles intersecting the μ axis perpendicularly.

Exercise 6.2 Verify that the construction of section 6.7.2 gives the (observed) expected Fisher information when applied to the (observed) expected yoke.

Exercise 6.3 Show that the Fisher information on $\mathcal{N}$ is invariant under $SL(2, \mathbf{R})$. Find a statistical interpretation of this fact and send it to the authors.

CHAPTER 7

Asymptotics

7.1 Asymptotics

7.1.1 *Introduction*

Many of the applications of differential geometry to statistics occur in the area of asymptotic theory. We first give an outline of the basic problem from a geometric point of view.

We have been interested, so far, in a family of distributions P on a sample space Ω with a base measure μ. Asymptotic theory is concerned with the effects of taking repeated larger samples from Ω. That means we replace Ω by Ω^N for some natural number N and the base measure by the product measure μ^N. Any probability measure μ on Ω defines a product measure μ^N on the product sample space Ω^N. If μ has density p with respect to some base measure then μ^N has density

$$p_N(x_1, \ldots, x_N) = p(x_1)p(x_2)\ldots p(x_N)$$

with respect to the product of the base measure. So for every N the space P is also a space of probability measures on Ω^N.

We can now consider families of estimators

$$u_N : \Omega^N \to P$$

and we are interested in their asymptotic behaviour. For instance we might like to know the limit of these estimators and develop an asymptotic expansion for them in inverse fractional powers of the square root of N.

An example of such an estimator is the mle

$$M_N : \Omega^N \to P$$

that we defined in section 6.7.1 or below in section 7.3.1.

If u_N was a random variable taking values in, say, $\mathbf{R}$ then we could analyse its mean, variance and higher moments. However, in the manifold case we cannot do this. Indeed it is not even possible to calculate anything as simple as a mean. We could choose co-ordinates and then try and analyse how the results change as we change the co-ordinates. This is partly what we need strings for in the later sections on higher-order differential geometry. However, it is more in keeping with the philosophy of manifolds that we have presented to choose any function f on P and consider the real-valued random variable $f \circ u_N$. This, of course, includes the case where we choose co-ordinates.

Before dealing with these questions we need to revise some basic notions from statistics, the mle and the definition of sufficient statistics.

7.2 Estimators

7.2.1 *Introduction: unbiased estimators*

A function $f : \Omega \to P$ is called an estimator. It tells us that if we choose a point x from Ω the true distribution is $f(x)$. Of course we hope to have a sample of larger size than 1 so we consider the measure spaces Ω^N formed by taking the N-fold product of Ω. Any measure ρ on Ω defines a product measures ρ^N on Ω^N. In particular a point $p\mu$ in P defines the measure $p_N \mu^N$ where

$$p_N(x_1, \ldots, x_N) = p(x_1)p(x_2)\ldots p(x_N)$$

In such a case we are interested in sequences of estimators $f_N : \Omega^N \to P$ and in particular in their asymptotic behaviour as the size of the sample (N) approaches infinity.

As we have defined it any function is an estimator and, of course, some functions will tell us nothing at all about the true point ! So we wish to consider particular estimators and particular properties a good estimator should have. One such property is the idea that the average over all samples of an estimate should be the true point. Such estimators are called *unbiased.* To give a precise definition we have to make sense of the idea of averaging points on a manifold P. This we will not do here but instead restrict attention, for now, to the case that P admits global co-ordinates. Then if $\varphi: P \to \mathbf{R}^n$ is a choice of co-ordinates we say that f is unbiased, with respect to the co-ordinates φ, if

$$E_p(\varphi \circ f) = \varphi(p)$$

Similarly we say that f_N is *asymptotically unbiased* if

$$\lim_{N \to \infty} E_p(\varphi \circ f_N) = \varphi(p)$$

7.3 The maximum likelihood estimator (mle)

One estimator that can be defined in a broad class of examples is the maximum likelihood estimator (mle). Consider for a fixed x in Ω the function $\ell(x)$ which is a function on P. If the mle is to be defined this must have a unique maximum. We will always assume that that maximum occurs in the interior of P, then, just as in multivariable calculus, if the maximum occurs at p we must have $d\ell(x)_p = 0$. This is readily checked by choosing local co-ordinates around the point p. The mle is the function M which assigns to the point x in Ω the point $M(x)$ in P at which this unique maximum occurs. So we have

$$M: \Omega \to P$$

and

$$d_{M(x)}\ell(x) = 0$$

Notice that M is a random variable with values in a manifold. This equation is usually called the likelihood equation. In co-ordinates θ it takes the form of the r equations

$$\frac{\partial \ell}{\partial \theta^i}(M(x)) = 0$$

As we noted before we are departing here from the usual conventions in statistics where the random variable $x \mapsto \theta^i(M(x))$ is denoted by $\hat{\theta}^i$ and no name is given to the mle.

Another way to view the mle is to consider the function

$$\Omega \times P \to \mathbf{R}$$
$$(x, p) \mapsto d_p(\ell)(x).$$

If the set where this function is zero is is the graph of (some other) function $M: \Omega \to P$ then this M is called the mle.

7.3.1 *The mle of an exponential family*

To find the mle of the exponential family we assume first that the likelihood has a unique maximum. General conditions for this are given in Barndorff-Nielsen (1978). Then we can find that maximum using calculus. We have

$$d_p\ell(x) = \sum_i x^i d_p\theta^i - d_pK$$

and expanding in the basis $d_p\theta^i$ we see that this is zero precisely when

$$x^i - \frac{\partial K}{\partial \theta^i}(p) = 0$$

for all $i = 1, \ldots, n$. So we must have

$$x^i - \frac{\partial K}{\partial \theta^i}(M(x)) = 0$$

hence the inverse of the mle is

$$M^{-1}(p) = (\frac{\partial K}{\partial \theta^1}(p), \ldots, \frac{\partial K}{\partial \theta^n}(p))$$

We shall call an exponential family *complete* if the mle is a bijection: again, general conditions for a family to be complete are given in Barndorff-Nielsen (1978). In such a case the inverse of the mle defines a set of co-ordinates on P called the *expectation co-ordinates*

$$\eta^i(p) = \frac{\partial K}{\partial \theta^i}(p) = E_p(x^i)$$

Let us prove that the mle for the exponential family is unbiased in the expectation co-ordinates. First note that $\eta^i \circ M = x^i$. Then we have that

$$E_p(\eta^i \circ M) = E_p(x^i) = \frac{\partial K}{\partial \theta^i}(p) = \eta^i(p)$$

Example 7.3.1 *The expectation co-ordinates of the exponential family.* We leave it as an exercise for the reader to show that the expectation co-ordinates for the normal family are

$$\eta^1 = \mu$$

and

$$\eta^2 = \sigma^2 - \mu^2$$

Example 7.3.2 *The mle for the normal family.* If we have a sample of size N then the likelihood is

$$p(x_1, \ldots, x_N) = \frac{1}{(\sqrt{2\pi}\sigma(p))^N} \prod_{i=1}^{N} \exp\left(\frac{-(x_i - \mu(p))^2)}{2\sigma(p)^2}\right)$$

and therefore the log-likelihood for a sample of size N is

$$\ell(p)(x_1, \ldots, x_N) = \sum_{i=1}^{N} \frac{-(x_i - \mu(p))^2}{2\sigma(p)^2} - N\log(\sqrt{2\pi}\sigma(p))$$

Using the μ and σ co-ordinates the likelihood equation becomes

$$\frac{\partial \ell}{\partial \mu}(M(x)) = \frac{-2}{\sigma(M(x))} \sum_{i=1}^{N} (x_i - \mu(M(x)))$$

and

$$\frac{\partial \ell}{\partial \sigma}(M(x)) = \sum_{i=1}^{N} \frac{(x_i - \mu(M(x)))^2}{\sigma(M(x))^3} - N\frac{1}{\sigma(M(x))}$$

These have the solutions

$$\hat{\mu}(x) = \mu(M(x)) = \frac{1}{N}\sum_{i=1}^{N} x_i = \bar{x}$$

and

$$\hat{\sigma}(x) = \sigma(M(x)) = \frac{1}{N^{\frac{1}{2}}}\left(\sum_{i=1}^{N}(x_i - \hat{\mu}(x))^2\right)^{\frac{1}{2}}$$

Notice that if $N = 1$ the solution has $\sigma(M(x)) = 0$ which is not in $\mathcal{N}$. This corresponds to the fact that the amongst all the normal densities of mean μ the height of the maximum increases as the $\sigma \to 0$ and the densities become delta measures at μ.

We see also that

$$E(\hat{\mu}) = \mu$$

and that

$$E(\hat{\sigma}^2) = \sigma^2 - \mu^2$$

so that the mle is not unbiased in the co-ordinates (μ, σ^2). Of course from the previous section we know that it is unbiased in the expectation co-ordinates which for the normal family are

$$\eta^1 = \mu \quad \text{and} \quad \eta^2 = \sigma^2 - \mu^2$$

Example 7.3.3 *The mle for a curved exponential family.* Assume that P is a complete exponential family. Consider a curved exponential family $Q \subset P$ and its mle

$$M_Q\colon P \to Q$$

We are interested in the fibres of this map. That is the sets

$$M_Q^{-1}(q) = \{x \mid M_Q(x) = q\}$$

If we apply the mle for Q to a point $(x^1, \ldots, x^n)$ then this is mapped to the point q such that

$$\sum_i x^i \frac{\partial \theta^i}{\partial \chi^j}(q) - \frac{\partial K}{\partial \chi^j}(p) = 0$$

This defines a subspace in the affine structure of Ω. So we conclude that the fibres of the mle of curved exponential family are affine subspaces of the sample space.

7.3.2 *The mixture affine structure*

It is clear from the formula

$$p(x, \theta) = \exp(\sum_i x^i \theta^i - K(\theta))$$

for a density in an exponential family that there is some kind of duality relating the x and θ variables or the points of Ω and the points of P. We show that this duality implies that if we denote the space of random variables, up to addition of constants that acts on P by V then Ω is an affine space for V^*.

Any two points p and q in an affine space define a vector, the vector that translates one point to another. It is convenient to denote this by $p - q$. In the case of an exponential family the vector translating p to q is the random variable

$$q - p = (\sum_i x^i \theta^i(q)) - (\sum_i x^i \theta^i(p))$$

up to addition of constants. If x and y are two points in Ω then we can define a vector $x - y$ in V^* by its action on V. We define

$$(x - y)(f) = f(x) - f(y)$$

The vectors in V are all of the form $\sum_i \theta^i x^i$ up to addition of constants so this can also be written

$$(x - y)(\sum_i \theta^i x^i) = \sum_i \theta^i x^i - \sum_i \theta^i y^i$$

In the case of an exponential family we can identify Ω and P with the mle and this becomes

$$\begin{aligned} p - q =& M^{-1}(p) - M^{-1}(q) \\ =& \sum_i \theta^i \eta^i(p) - \sum_i \theta^i \eta^i(q) \end{aligned}$$

or the affine space structure for which the expectation co-ordinates are affine co-ordinates.

When it is necessary to distinguish between these two affine structure we shall call the one that P inherits from $\mathcal{P}$ the expectation affine strucure and this new one the mixture affine structure.

Notice the potentially confusing situation that the expectation co-ordinates are *not* the affine co-ordinates for the expectation affine structure. The reason for calling this the mixture affine structure is that if we form a mixture of two probability measures

$$tp + (1-t)q$$

in an exponential family then

$$E_{tp+(1-t)q}(x^i) = tE_p(x^i) + (1-t)E_q(x^i)$$

which as t varies is a line in the mixture affine structure.

7.4 The Cramer–Rao inequality

Assume that $\theta\colon P \to \mathbf{R}^n$ are global co-ordinates and that $u\colon \Omega \to P$ is an unbiased estimator. So we have

$$E_p(\theta \circ u) = \theta(p)$$

and differentiating this gives

$$\delta^i_j = E_p(\frac{\partial \ell}{\partial \theta^j}\theta^i) \tag{7.4.1}$$

We wish to show that the covariance matrix for the random variable $\theta^i \circ u$ is greater than the inverse of the Fisher information matrix in the sense that

$$\mathrm{cov}(\theta^i \circ u, \theta^j \circ u) - g^{ij} \tag{7.4.2}$$

is positive semi-definite. The inequality (7.2.2) is called the Cramer–Rao inequality. We note that

$$E_p((\theta^i \circ u - E_p(\theta^i \circ u) - \frac{\partial \ell}{\partial \theta^k}g^{ik})(\theta^j \circ u - E_p(\theta^j \circ u) - \frac{\partial \ell}{\partial \theta^l}g^{jl})$$

is positive semi-definite, where we employ the Einstein summation convention so that k and l are summed over.

Expanding this gives

$$\mathrm{cov}(\theta^i \circ u, \theta^j \circ u) - 2E_p((\theta^i \circ u - E_p(\theta^i \circ u))\frac{\partial \ell}{\partial \theta^l}g^{jl}) + g^{ij}$$

and applying (7.2.1) gives

$$\begin{aligned} E_p((\theta^i \circ u - E_p(\theta^i \circ u))\frac{\partial \ell}{\partial \theta^l} g^{jl}) =& E_p((\theta^i \circ u)\frac{\partial \ell}{\partial \theta^l} g^{jl}) \\ =& g^{ij} \end{aligned}$$

which yields the desired result.

Example 7.4.1 *The mle of an exponential family attains the Cramer–Rao lower bound.* We show that for an exponential family with the expectation co-ordinates $\eta^a(p) = E_p(x^a)$ the mle attains the Cramer–Rao lower bound. Denote by θ^i the canonical co-ordinates. Note that

$$\frac{\partial \eta^a}{\partial \theta^i} = \frac{\partial K}{\partial \theta^a \partial \theta^i} = -g_{ai}$$

so that

$$g^{ab} = -\frac{\partial^2 K}{\partial \theta^a \partial \theta^b} = -\frac{\partial \eta^a}{\partial \theta^i}\delta^i_b = g_{ab}$$

The reader will notice that in this case there is a confusion with the placing of indices and a grave danger of confusing g^{ab} and g_{ab}.

As the mle is the inverse of the expectation co-ordinates we have

$$\begin{aligned} \mathrm{cov}(\eta^a \circ M, \eta^b \circ M) =& \mathrm{cov}(x^a, x^b) \\ =& E((x^a - E(x^a))(x^b - E(x^b)) \\ =& E((x^a - \frac{\partial K}{\partial \theta^a})(x^b - \frac{\partial K}{\partial \theta^b})) \\ =& E(\frac{\partial \ell}{\partial \theta^a}\frac{\partial \ell}{\partial \theta^b}) \\ =& g_{ab} \\ =& g^{ab} \end{aligned}$$

which shows that in the canonical co-ordinates the mle for the exponential family attains the Cramer–Rao lower bound.

It is not clear from the way that we have formulated it what the invariance properties of the Cramer–Rao lower bound are. Let us reformulate it in an invariant way that will motivate the later definition of co-ordinate strings.

We replace the choice of co-ordinates θ by a smooth function

$$\psi_p : P \to T_p P$$

which is a diffeomorphism in an open set about p and satisfies $\psi(p) = 0$. We do this smoothly for every p. Then we say that an estimator

$$u : \Omega \to P$$

is unbiased relative to ψ if for all p we have

$$E_p(\psi_p) = 0$$

To relate this to the previous discussion let θ be co-ordinates and define

$$\psi_p(q) = \sum_i (\theta^i(q) - \theta^i(p)) \frac{\partial}{\partial \theta^i}(p)$$

Then if $\psi_p(q)$ is unbiased if and only if

$$E_p(\theta^i \circ u) = \theta^i(p)$$

Given the estimator u we define an inner product $g_{u,\psi}$ on the cotangent space, $T_p^* P$, at each p by

$$g_{u,\psi}(\omega, \rho) = E_p(\omega(\psi_p \circ u)\rho(\psi_p \circ u))$$

for any ω and ρ in $T_p^* P$. Notice that if we have co-ordinates θ then

$$\begin{aligned} g_{u,\psi}(d\theta^i, d\theta^j) &= E_p((\theta^i \circ u - \theta^i(p))(\theta^j(q) \circ u - \theta^j(p)) \\ &= \operatorname{cov}(\theta^i \circ u, \theta^j \circ u) \end{aligned}$$

The proof we have given above immediately generalises to give a proof that

$$g_{u,\psi} \geq g$$

where g is the Fisher information inner product on $T_p^* P$. Again the inequality here is in the sense that the difference is positive semi-definite.

Objects similar to this ψ will be introduced in Chapter 9 and called co-ordinate strings.

7.5 Sufficiency

Although we focus attention on P, our definition of a statistical manifold is really a pair (Ω, P) consisting of a measure space Ω and a manifold of probability distributions on Ω. Statisticians often wish to know whether they need to consider all the information in the sample space Ω. For instance if Ω is the N-fold product of another sample space then often it is enough to focus attention on the mean of the N points. Mathematically we describe this idea as a map

$$Y : \Omega \to \Omega'$$

such that if $Y(\omega) = Y(\rho)$ then ω and ρ 'contain the same information about the true distribution'. The notion of sufficiency makes precise this idea of not losing information. Before coming to it we have to consider some properties of such maps.

7.5.1 *Conditional distributions*

Let Ω be a probability space with probability measure ρ. If $Y : \Omega \to \Omega'$ is a map then we can construct the pushout measure $Y_*\rho$ on the space Ω' by defining a subset $A \subset \Omega'$ to be measurable if $Y^{-1}(A)$ is measurable and then defining its measure to be

$$Y_*(\rho)(A) = \rho(Y^{-1}(A))$$

Similarly if X is a random variable on Ω then we define $E(X|Y)$ a random variable on Ω by

$$\int_{Y^{-1}(A)} X\rho = \int_A E(X|Y)\rho$$

for any $A \subset \Omega'$.

In other words $E(\ |Y)$ is a map

$$E(\ ,|Y) : R_\Omega \to R_{\Omega'}$$

from random variables on Ω to random variables on Ω'.

Recall that under quite general conditions Barndorff-Nielsen, Blæsild and Eriksen (1989) if we have a measure space Ω and a map $Y : \Omega \to \Omega'$ we can find a unique family of measures $\{v_x(\mu)_x | x \in \Omega'\}$

where $v_x(\mu)_x$ is a measure on $Y^{-1}(x)$ such that for any measurable function f on Ω we have

$$\int_{\Omega} f\mu = \int_{\Omega'} (\int_{Y^{-1}(x)} f v_x(\mu)) Y_*(\mu)$$

that is we can integrate f on Ω by first integrating over all the fibres of Y and then integrating the resulting function on Ω' against the pushout of μ. The notation $\nu(\mu)$ is supposed to suggest the vertical part of μ on the fibre over x. There is also, of course, a dependence on Y which we have suppressed from the notation.

If we perform such a decomposition then we have

$$E(X|Y)(x) = \int_{Y^{-1}(x)} X v_x(\mu)$$

It is useful to know how these ideas fit together with the affine structure of $\mathcal{M}$. Let p be a probability measure on Ω and f a function in R_Ω. Then we have for any subset A of Ω'

$$\begin{aligned} Y_*(e^f p)(A) &= \int_{Y^{-1}(A)} e^f p \\ &= \int_A (\int_{Y^{-1}(x)} e^f v_x(p)) Y_*(p) \\ &= [(\int_{Y^{-1}(x)} e^f v_x(p)) Y_*(p)](A) \end{aligned}$$

so that

$$Y_*(e^f p) = (\int_{Y^{-1}(x)} e^f v_x(p)) Y_*(p)$$

So for any g on Ω

$$\begin{aligned} \int_{\Omega} (ge^f) p &= \int_{\Omega'} (\int_{Y^{-1}(x)} (ge^f) v_x(p)) Y_*(p) \\ &= \int_{\Omega'} \int_{Y^{-1}(x)} g \left\{ \frac{e^f}{\int_{Y^{-1}(x)} e^f v_x(p)} v_x(p) \right\} Y_*(p) \end{aligned}$$

Because of the uniqueness of the measures $v_x(\mu)_x$ we must have

$$v_x(e^f p) = \frac{e^f}{\int_{Y^{-1}(x)} e^f v_x(p)} v_x(p)$$

7.5.2 *Sufficient statistics: the factorisation principle.*

Consider now a statistical manifold P and a random variable

$$Y : \Omega \to \Omega'$$

as above. If we think of P as a family of measures on Ω then for each p we have the vertical measures $v_x(p\mu)$. The statistic Y is called *sufficient* if the $v_x(p\mu)$ do not depend on p. The point of the definition is that we can think of P as a family of measures on Ω' and recover the family of measures on Ω as long as we know what $v_x(p\mu)$ is. We say that (P, Ω') is a sufficient reduction of (P, Ω).

Often it will not be clear what this decomposition into vertical measures looks like and it is more useful to use the *factorisation principal.* If we have fixed an origin μ then we have seen that

$$v_x(p\mu) = \frac{p}{\int_{Y^{-1}(x)} p v_x(\mu)} v_x(\mu)$$

If we define

$$q : P \times \Omega' \to \mathbf{R}$$

by

$$q(p, x) = \int_{Y^{-1}(x)} p v_x(\mu)$$

and

$$r(x) = \frac{p}{\int_{Y^{-1}(x)} p v_x(\mu)}$$

then we see that the density p factorises as

$$p(x) = q(p, Y(x)) r(x)$$

On the other hand assume that we choose an origin μ and we can factorise the densities in P in this form for some q and r. Then we have

$$\begin{aligned} v_x(p\mu) &= \frac{p}{\int_{Y^{-1}(x)} p v_x(\mu)} v_x(\mu) \\ &= \frac{q(p, Y(x)) r(x)}{\int_{Y^{-1}(x)} q(p, Y(x)) r(x) v_x(\mu)} v_x(\mu) \\ &= \frac{r(x)}{\int_{Y^{-1}(x)} r(x) v_x(\mu)} v_x(\mu) \end{aligned}$$

which is clearly independent of p. So Y being sufficient for the family P is equivalent to being able to factorise the densities as

$$p(x) = q(p, Y(x))r(x)$$

Let us consider some of the properties of a sufficient reduction. We have that

$$v_x(e^f p) = \frac{e^f}{\int_{Y^{-1}(x)} e^f v_x(p)} v_x(p)$$

This is a general fact. If, in addition, the statistic Y is sufficient for the family consisting of $e^f p$ then it follows that

$$e^f = \int_{Y^{-1}(x)} e^f v_x(p)$$

or that f is constant on the fibres of the statistic, i.e. on the sets $Y^{-1}(x)$.

The functions constant on the fibres of Y form a vector subspace of R_Ω which we shall denote by R_Y. It is clearly isomorphic to $R_{\Omega'}$. If we let $\mathcal{M}_Y$ be the subaffine space of $\mathcal{M}$ containing $P.R_Y$ then many of the constructions we shall perform in the remainder of this book will only depend on P and its embedding into $\mathcal{M}_Y$. In particular the tangent space to $\mathcal{P}$ is contained in R_Y so the score takes its values in R_Y. So we would lose nothing if we replaced R_Ω by R_Y and M_Ω by $\mathcal{M}_Y$. From a statistical viewpoint we are just stating the well-known fact that if Y is sufficient for the family P then inference depends on the data in Ω only through Y.

Pushing down measures defines a map

$$Y_*: \mathcal{P}_\Omega \to \mathcal{P}_{\Omega'}$$

Because $v_x(p)$ is a probability measure on $Y^{-1}(x)$ it follows if f is in R_Y that

$$Y_*(e^f p) = e^f Y_*(p)$$

that is

$$Y_*(p + f) = Y_*(p) + f$$

Using this formulae it is easy to show that

$$Y_*: \mathcal{M}_Y \to \mathcal{M}_{\Omega'}$$

is injective and hence is an isomorphism of affine spaces from $\mathcal{M}_Y$ to the image of Y_* inside $\mathcal{M}_{\Omega'}$.

7.5.3 *Minimal sufficiency*

If we compose a sufficient statistic $Y: \Omega \to \Omega'$ with any bijective map it is easy to check that the result is also a sufficient statistic. This suggests that what is important in the definition of sufficiency is the partitition of the sample space into the fibres of Y. That is the partition

$$\Omega = \cup\{Y^{-1}(x) | x \in \Omega'\}$$

Indeed the theorem on decomposition of measures we described above applies in this case also. Consider a partition Π of Ω. That is Π is a collection of disjoint subsets of Ω such that

$$\Omega = \cup_{A \in \Pi} A$$

If x is in Ω we shall use the notation $[x]$ for the (unique!) element of Π containing x. The projection $\pi: \Omega \to \Pi$ is defined by $x \mapsto [x]$. A measure ρ on Ω can be decomposed into family of measures $v_{[x]}(\rho)$ on the subsets $[x]$ and a pushdown measure $\pi_*(\rho)$ on the space Π. In the case of a statistic Y the partition is just the fibres of the map. We call a partition sufficient for the family P if the measures $v_{[x]}(p\mu)$ are independent of p. We can order partitions by refinement. A partition Π' is said to be smaller that a partition $\mathcal{P}$ if the sets in $\mathcal{P}'$ are obtained from the sets in Π by breaking them up into subsets. As forming a sufficient partition of the sample space loses no information about the family but reduces the size of the sample space it would be nice if we could do this as much as possible. This means that we would like to find an equivalence relation which is minimal amongst all sufficient ones in the sense that any other sufficient statistic is obtained by subdividing the minimal one. The minimal statistic would then represent the maximal amount of clumping together of data that is possible without losinq information. Note that it is not at all obvious that we can do this. It is remarkable that we can and that moreover the minimal partition has a natural description.

7.5.4 *The likelihood ratio statistic.*

Let us define a partition of Ω, called the *likelihood ratio partition*, or *likelihood ratio statistic*, as follows. Define x and y to be in the same set of the partition if for all p in P we have that

$$\ell(x, p) = \ell(y, p) + c(x, y)$$

for some function $c(x, y)$. In terms of probability densities this means that

$$p(x) = p(y) \exp c(x, y)$$

If we denote by $\Gamma(T^*(P))$ the set of all sections of the cotangent bundle of P we see that the likelihood ration partition can be described as the fibres of the map

$$\begin{aligned} \Omega &\to \Gamma(T^*(P)) \\ x &\mapsto d\ell(x). \end{aligned}$$

Alternatively fix a point q in P, then the likelihood ratio partition is also the fibres of the map

$$\begin{aligned} \Omega &\to C^\infty(P, \mathbf{R}) \\ x &\mapsto (p \mapsto \tfrac{p(x)}{q(x)}) \end{aligned}$$

This map is called the likelihood ratio statistic.

Let us prove that this is a sufficient partition. Decompose the base measure μ according to the partition. If the partition is to be sufficient we need

$$\frac{p}{\int_{[x]} p v_{[x]}} v_{[x]} \tag{7.5.1}$$

to be independent of p for all x. If we fix an x in the equivalence class and let $c(x)$ be the function $y \mapsto c(x, y)$ we have

$$p = p(x) \exp c(x)$$

If we substitute this into (7.5.1) we obtain

$$\frac{c(x)}{\int_{[x]} c(x) v_{[x]}} v_{[x]}$$

which is clearly independent of p.

To prove that this is minimal let Π be a sufficient partition and $\pi \colon \Omega \to \Pi$ the projection. Let ℓ be the log-likelihood constructed from Ω and $\tilde{\ell}$ that constructed from Π. As for the case of statistics sufficiency implies that

$$\ell(x) = \tilde{\ell}([x]) + C(x)$$

for some random variable C. But then we have for any p

$$d_p\ell(x) = d_p\tilde{\ell}([x])$$

so that if $[x] = [y]$ we must have $d_p\ell(x) = d_p\ell(y)$.

Example 7.5.1 *Sufficiency of the mean for exponential families.* Consider a complete exponential family P on a sample space $\Omega = \mathbf{R}^r$. Then by defining product measures Barndorff-Nielsen, Blæsild and Eriksen (1989) we can consider the measures in P as measures on the sample space Ω^N. If we choose an origin μ the product densities are

$$p^N(x_{(1)}, \ldots, x_{(N)}) = \exp(\sum_{a=1}^{N}\sum_{i=1}^{r} x^i_{(a)}\theta^i - K(p))$$

If we define a map

$$Y: \Omega^N \to \Omega$$

by

$$Y(x_{(1)}, \ldots, x_{(N)}) = \frac{1}{N}\sum_{a=1}^{N} x_{(a)}$$

Then we have

$$p^N(x_{(1)}, \ldots, x_{(N)}) = \exp(N(\sum_{i=1}^{r} Y(x)^i\theta^i - K(p))$$

so that by the factorisation principle we see that Y is a sufficient statistic for the exponential family.

It is worth noting that if P is a complete family for Ω then Ω is an affine space. In such a case it is possible to define Y without identifying Ω with $\mathbf{R}^n$. We define $Y(x_{(1)}, \ldots, x_{(N)})$ to be the unique point in Ω such that

$$\sum_{a=1}^{N}(Y(x_{(1)}, \ldots, x_{(N)}) - x_{(a)}) = 0$$

7.6 The central limit theorem and the score form

The classical limit theorems are the canonical examples of the phenomena in which we are interested, so we consider them first. Assume that we have a sample space Ω with probability measure ρ. Consider a random variable X with values in a vector space V. Then this defines a linear map

$$\begin{aligned} X^*: \qquad V^* &\to R_\Omega \\ \xi &\mapsto (x \mapsto \xi(X(x))) \end{aligned}$$

which maps V^* into some subspace of R_Ω. If we have $V = \mathbf{R}^n$ and let ξ^i be the ith projection defined by $\xi^i((v^1, \ldots, v^n)) = v^i$ then this map sends ξ^i to X^i.

We can define a bilinear form on V^* by

$$\Sigma(\xi, \eta) = E(X^*(\xi)X^*(\eta))$$

This is called the variance of X.

Recall that if we have a positive definite inner product on V^* then this defines an isomorphism $V \simeq V^*$ and therefore a positive definite inner product on V. If Σ is positive definite we denote by Σ^* the inner product induced on V.

Assume now that we have a sequence of random variables X_i with mean μ and variance Σ. Let $\bar{X} = 1/N \sum_{i=1}^N X_i$ where we suppress the dependence on N. Then we have two standard limit results.

The weak law of large numbers tells us that the limit of the sequence

$$\bar{X}_*(\rho)$$

of measures on P is a delta measure at μ.

The central limit theorem tells us that if n is the dimension of V the limit of the sequence of measures

$$(\sqrt{N}\bar{X})_*(\rho)$$

is the measure

$$N(0, \Sigma^*) = \frac{1}{(2\pi)^{n/2}} \exp(\Sigma^*(x, x))\mu^*_\Sigma)$$

where μ_{Σ^*} is the density on V determined by the inner product Σ^*.

If V is the vector space $\mathbf{R}^n$ then $\Sigma = (\Sigma_{ij})$ is a positive definite matrix and this becomes

$$\frac{1}{(2\pi)^{n/2}} \exp(\sum_{i,j} x^i \Sigma_{ij} x^j) | \det(\Sigma_{ij})|^{-1/2} dx^1 \ldots dx^n$$

We can apply these results to the score form considered as a random variable

$$\begin{aligned} d_p(\ell) &: \Omega \to T_p^*(P) \\ x &\mapsto d_p(\ell)(x). \end{aligned}$$

For each N we also have

$$\begin{aligned} d_p(\ell_N) &: \Omega^N \to T_p^*(P) \\ x &\mapsto d_p(\ell)(x). \end{aligned}$$

and because of the additivity of the logarithm

$$d_p \ell_N(x_1, \ldots, x_N) = \Sigma_{i=1}^N d_p \ell(x_i)$$

It follows then that

$$\sqrt{N} \bar{d}_p(\ell) = \frac{1}{\sqrt{N}} d_p \ell_N$$

The induced map

$$d_p(\ell)(x)^* : T_p(P) \to R_\Omega$$

is the map that sends v in $T_p(P)$ to the random variable $x \mapsto \langle d_p(\ell)(x), v \rangle = v(\ell)(x)$. Hence it is the just the score map restricted to $T_p(P)$ and it follows that the variance of the score form is the Fisher-information metric as an inner product on $T_p^{**} = T_p P$. The central limit theorem then tells us that

$$\lim_{N \to \infty} (\frac{1}{\sqrt{N}} d_p \ell_N)_* (p\mu) = N(0, g_p^*)$$

7.7 Asymptotics of the mle

7.7.1 *The asymptotics of the unnormalised mle*

Let p be a distribution in P and consider the sequence of measures

$$M_{N*}(p\mu)$$

on P. It can be shown under fairly general conditions that the limit of these measures is a delta distribution supported at p. Interestingly the original proof of this due to Wald (1949) uses no co-ordinates on P. Indeed P does not even need to be a manifold only a metric space. As we have noted in Chapter 6 the Fisher information metric makes P into a metric space. However for the purposes of Wald's proof any (topological) metric on P would do.

We can think of this result as analogous to a law of large numbers for the mle.

7.7.2 *The asymptotics of the mle*

The result from the previous section shows that for large enough N nearly all the mass of the measure $M_{N*}(p\mu)$ is in the domain of some set of local co-ordinates about p. We can then choose a map

$$\psi : P \to T_p P$$

which restricted to an open set about P is a diffeomorphism. This we do by identifying $\mathbf{R}^n$ and $T_p P$ and then extending the co-ordinates to functions on all of P. We could work just in a co-ordinate chart around p but it is simpler to do it this way. By analogy with the central limit theorem we now consider

$$\sqrt{N}\psi : P \to T_p P$$

This is the map obtained by composing ψ with multiplication by $\sqrt{N}$.

We now consider the sequence of measures

$$(\sqrt{N}\psi)_* M_{N*}(p\mu)$$

on $T_p P$. We want to show that this is asymptotically the measure

$$N(0, g_p)$$

We proceed as in Cox and Hinkley (1974). If we choose a basis for T_pP then the map ψ becomes co-ordinates ψ^i and we can Taylor expand the likelihood equation about the point p. This gives us

$$0 = \frac{\partial \ell_N}{\partial \psi^i}(p) + \frac{\partial^2 \ell_N}{\partial \psi^i \partial \psi^j}(p)\psi^j \circ M_N + \dots$$

for all i. Let us denote by $\mathrm{Hess}_{\psi,p}(\ell_N)$ the matrix of second derivatives

$$\frac{\partial^2 \ell_N}{\partial \psi^i \partial \psi^j}(p)$$

and note that it depends on the choice of co-ordinates ψ. We can also regard $\mathrm{Hess}_{\psi,p}(\ell_N)$ as an inner product on T_pP whose components on the basis we have chosen form the matrix of second derivatives.

Then we have the approximate equation

$$d_p(\ell_N) = -\mathrm{Hess}_{\psi,p}(\ell_N)(\psi \circ M_N) + \dots. \tag{7.7.1}$$

Here we think of the inner product as mapping T_pP to T_p^*P.

The random variable $\mathrm{Hess}_{\psi,p}(\ell_N)$ equals $N\overline{\mathrm{Hess}_p(\ell)}$ so the law of large numbers tells us that

$$\frac{1}{N}\mathrm{Hess}_{\psi,p}(\ell_N)$$

is asymptotic to $E(\mathrm{Hess}_p(\ell)) = -g_p$. If we assume that the terms in equation (7.7.1) buried in the dots have some faster asymptotic decay in $1/N$ then we can deduce that the limit of

$$(\sqrt{N}\psi)_* M_{N*}(p\mu)$$

is

$$N(0, g_p)$$

To convert this into a rigorous argument we only need to use Taylor's theorem and assume some kind of bound on the remainder. Note also that the limiting measure on T_pP is independent of the choice of ψ.

Example 7.7.1 *The unit normal family.* It it worth considering one example where we can calculate all these things exactly. Let

P be the family $\mathcal{N}(\mathbf{R}, 1)$ of normal distributions with constant variance matrix equal to the identity. The sample space is $\Omega = \mathbf{R}$ and P is the set of measures

$$p(\mu)(x) = \frac{1}{(2\pi)^{1/2}} \exp(-\frac{1}{2}|x - \mu|^2)dx$$

We identify P with $\mathbf{R}$ by identifying each distribution with its mean. The product sample space is then $\mathbf{R}^N$ and the mle is easily calculated to be

$$M_N(x^1, \ldots, x^N) = \frac{1}{N} \sum_{i=1}^{N} x^i$$

The product measure on $\mathbf{R}^N$ is again a Gaussian

$$\frac{1}{(2\pi)^{N/2}} \prod_{i=1}^{N} \exp(-\frac{1}{2}|x^i \mu|^2)dx^1 \ldots dx^N$$

Let us assume that $p = 0$ and that ψ is just the usual co-ordinates on $\mathbf{R}$. To calculate the pushdown measure under M^N we introduce new co-ordinates on $\mathbf{R}^N$ by choosing a basis

$$\begin{aligned} e_1 =& \frac{1}{\sqrt{N}}(1, 0, \ldots, 0, 1), \\ e_2 =& \frac{1}{\sqrt{2}}(1, -1, 0, \ldots, 0) \\ & \vdots \\ e_N =& \frac{1}{\sqrt{2}}(0, \ldots, 0, 1, -1) \end{aligned}$$

We can now define co-ordinates $y^1, \ldots, y^N$ by using this basis. The important thing to note is that this is an orthonormal basis and therefore the product measure in these co-ordinates is still a Gaussian

$$\frac{1}{(2\pi)^{N/2}} \prod_{i=1}^{N} \exp(-\frac{1}{2}|y^i|^2)dy^1 \ldots dy^N$$

In these new co-ordinates the mle is just

$$M_N(y^1, \ldots, y^N) = \frac{1}{\sqrt{N}} y^1$$

so that calculating the pushdown measure is just a matter of integrating out the remaining variables and finding that

$$\sqrt{N}\psi({}_*M_{N*}(p\mu) = \frac{1}{(2\pi)}\exp(-\frac{1}{2}|x-\mu|^2)dx$$

In the case of the unit normal family the asymptotic result is therefore exact. One of the aims of the application of differential geometry is to replace $\mathcal{N}(\mathbf{R}, 1)$ by a general statistical manifold $\mathcal{P}$ and replace this exact result by an asymptotic formulae in powers of $1/\sqrt{N}$ the higher terms of which should be related to the way in which the family curves inside $\mathcal{P}$.

7.8 Asymptotics and exponential families

We review some results from Amari (1985). Consider an exponential family P on $\Omega = \mathbf{R}^n$. We need to observe two important facts. The first is that

$$p_N(x_1,\ldots,x_N) = p(x_1)p(x_2)\ldots p(x_N) = p^N(m_M(x_1,\ldots,x_N)$$

where

$$m_N(x_1,\ldots,x_N) = \frac{1}{N}\sum_{i=1}^{N} x_i$$

is the mean and $p^N(y)$ denotes the Nth power of $p(y)$.

Consider the map $m_N\colon\Omega^N \to \Omega$. Recall the definition of the pushout of a measure given in section 7.5.1. Applied in this case it follows that

$$m_{N*}(p_N\mu^N) = p^N(m_{N*}\mu^N).$$

The second important fact we need is that

$$M \circ m_N = M_N.$$

Indeed if $(x_1,\ldots,x_N)$ are observations from Ω^N their maximum likelihood estimate occurs where

$$\sum_i x_i^j + N\frac{\partial K}{\partial\theta^j} = 0$$

for all j. On the other hand if we take the mean and then the maximum likelihood estimate of that it occurs at the point where

$$\frac{1}{N}\sum x_i^j + \frac{\partial K}{\partial \theta^j} = 0$$

Clearly these are the same point!

What this means for asymptotics is that if f is a function on P and we want to calculate

$$E(f \circ M_N) = \int_{\Omega^N} f \circ M_N p_N \mu^N$$

then this is the same as

$$\int_{\Omega^N} f \circ M_N p_N \mu^N = \int_{\Omega^N} f \circ M_N \circ m p_N \mu^N$$

or

$$\int_{\Omega} f \circ M p^N m_{N*} \mu^N$$

In all the calculations we do we shall see that it is possible to avoid the fact that we don't know much about $m_{N*}\mu^N$. For the remainder of this section by $E(f)$ we shall mean

$$\int_{\Omega} f \circ M p^N m_{N*} \mu^N$$

so the dependence on N and p is implicit.

We shall assume that the point p has expectation co-ordinates equal to 0. That means that

$$\frac{\partial K}{\partial \theta^i}(p) = 0$$

The function f can be Taylor expanded about p using the expectation co-ordinates as

$$f = \sum_{\alpha} \frac{1}{\alpha!}[D_\eta^\alpha(f)]\,|_{\eta=0}\eta^\alpha$$

where we adopt the usual multi-index notation

$$\alpha = (\alpha_1, \ldots, \alpha_n), \alpha! = \alpha_1! \ldots \alpha_n!,$$

$$|\alpha| = \alpha_1 + \ldots + \alpha_n$$

and

$$D_\eta^\alpha = \left(\frac{\partial}{\partial \eta^1}\right)^{\alpha_1} \ldots \left(\frac{\partial}{\partial \eta^n}\right)^{\alpha_n}$$

By composing with M we obtain an expansion for $f \circ M$ in terms of the random variables $x^\alpha = (x^1)^{\alpha_1} \ldots (x^n)^{\alpha_n}$ where the x^i are the co-ordinates on Ω in the definition of the exponential family. Taking expectations of both sides gives

$$E(f) = \sum_\alpha \frac{1}{\alpha!}(D_\eta^\alpha f)|_{\eta=0} E(\bar{x}^\alpha)$$

The pushout of the measure $p^N m_{N*}\mu^N$ is a probability measure so

$$\exp NK(\theta) = \int_\Omega \exp\left(\sum_{i=1}^n N x^i \theta_i\right) m_{N*}\mu^N$$

Applying D_θ^α to both sides we obtain

$$E(\bar{x}_N^\alpha) = \frac{1}{N^{|\alpha|}} \exp(-NK(\theta)) D_\theta^\alpha (\exp NK(\theta))$$

and finally

$$E(f) = \sum_\alpha \frac{1}{\alpha!} \frac{1}{N^{|\alpha|}} (D_\eta^\alpha f)(D_\theta^\alpha \exp NK(\theta)) \exp -NK(\theta)$$

We can rewrite the results of Amari as follows. If we collect together the derivatives of order $k = |\alpha|$ we obtain the expression

$$E(f) = \sum_k \frac{1}{k!} \frac{1}{N^k} g\big(\nabla_M^{(k)} f, \exp(-NK) \nabla_E^{(k)} \exp(NK)\big)$$

Here we denote by

$$\nabla_M^{(k)}(f)$$

the k-linear symmetric function of tangent vectors whose value on the expectation co-ordinate basis is

$$\nabla_M^{(k)}(f)\left(\frac{\partial}{\partial \eta^1}, \ldots, \frac{\partial}{\partial \eta^k}\right) = \frac{\partial^k f}{\partial \eta^1 \cdots \partial \eta^k}$$

and by $\nabla_E^{(k)}(f)$ the same object with the η co-ordinates replaced by the θ co-ordinates.

Expanding in N up to terms including those of order $1/N^2$ we obtain

$$E(f) = f(p) + \frac{1}{2N} g(\nabla_M^{(2)} f, \nabla_E^{(2)} K)(p) + \frac{1}{N^2}\left(\frac{1}{3!} g(\nabla_M^{(3)} f, \nabla_E^{(3)} K)(p) + g(\nabla_M^{(4)} f, \nabla_E^{(2)} K \nabla_E^{(2)} K)(p)\right) + \ldots$$

Notice that we have not been able to obtain an asympotic formula for the measure but for its expectations which is, of course, equivalent information. As we indicated at the end of the previous section this expansion does have higher terms.

Amari defines the bias corrected f^* defined by

$$f^* = \sqrt{N}\left(f - \frac{1}{2N} g(\nabla_M^2 f, \nabla_E^2 K)\right)$$

If f, h are two functions, after a couple of pages of calculations and some suprising cancellations we obtain, up to and including terms of order $1/N$,

$$E(f^* h^*) = g(\nabla_M f, \nabla_M h) + \frac{1}{2N} g(\nabla_M^{(2)} f, \nabla_M^{(2)} h) + \ldots$$

7.8.1 *Curved exponential families*

Assume now that we have an estimator $u: P \to Q \subset P$ for a curved exponential family Q. Let $U_p = u^{-1}(p)$. If we restrict to functions of the form $f = u \circ f', f': X \to \mathbf{R}$ we obtain the results of Amari. By decomposing T_pP as $T_pP = T_pU_p \oplus T_pQ$ we obtain similarly

$$S^2(T_p^* P) = S^2(T_p^* Q) \oplus T_p^* U_p \otimes T_p^* P \oplus S^2(T_p^* U_p)$$

and accordingly let

$$\nabla_M^{(2)} f = \nabla_M^{(2,0)} f + \nabla_M^{(1,1)} f + \nabla_M^{(0,2)} f$$

Here we are denoting by $S^2(V^*)$ the space of all symmetric, bilinear functions on a vector space V. Amari shows that for a first-order efficient estimator, T_pU_p and T_pP are orthogonal, so we obtain

$$g(\nabla_M^{(2)} f, \nabla_M^{(2)} h) = g(\nabla_M^{(2,0)} f, \nabla_M^{(2,0)} h) + 2g(\nabla_M^{(1,1)} f, \nabla_M^{(1,1)} h) + g(\nabla_M^{(0,2)} f, \nabla_M^{(0,2)} h)$$

We are interested in the three terms in this expression. The first is

$$g(\nabla_M^2 f, \nabla_M^2 h)$$

where g is the Fisher information metric and ∇_M the mixture connection. This term depends only on the geometry of Q and is independent of the choice of u.

After some calculations we can put the second term in the following form. We first introduce $g(df, \alpha)$, a one-form with values in the normal space at p, by defining it in co-ordinates as

$$g(df, \alpha)_i = \alpha_{ij} \frac{\partial f}{\partial \theta^k} g^{jk}$$

where α is the second fundamental form defined in Chapter 1. Then

$$g(\nabla_M^{(1,1)} f, \nabla_M^{(1,1)} h) = g^{ij} \langle g(df, \alpha)_i, g(dh, \alpha)_j \rangle. \qquad 6.17$$

This term clearly vanishes if the second fundamental form vanishes at p.

To understand the form of the last term we by define α^m to be the mixture second fundamental form for $U_p = u^{-1}(p)$ the fibre of u at p. If ϕ^a are co-ordinates on U_p then α^m_{ab} is in the normal space to U_p at p, that is the tangent space to P at p so we can pair it with df to obtain a number (α^m_{ab}, df). Then we have

$$g(\nabla_M^{(0,2)} f, \nabla_M^{(0,2)} h) = (\alpha^m_{ab}, df)(\alpha^m_{cd}, dh) g^{ac} g^{bd}$$

So this term vanishes if α^m vanishes at p.

Notice that if Q is exponential and u is the maximum likelihood estimator then Q is expectation affine and U_p is mixture affine so both B and C vanish. On the other hand if Q is fixed then if we want to minimize the size of this term then we need to choose the estimator u so that where its fibres U_p intersect Q the mixture second fundamental form for U_p in P vanishes. An estimator that minimises this term is called third order efficient and we have thereby deduced a condition for third order efficiency.

These calculations are taken from Amari (1985). To relate our notation to his take f and h to be co-ordinate functions for co-ordinates u^a and u^b then define:

$$(\Gamma^m)^{2ab} = g(\nabla^2_M u^a, \nabla^2_M u^b)$$

$$(H^e_M)^{2ab} = g(\nabla^{(1,1)}_M u^a, \nabla^{(1,1)}_M u^b) = g^{ij}\langle g(du^a,\alpha)_i, g(du^b,\alpha)_j\rangle$$

and

$$(H^m_A)^{2ab} = g(\nabla^{(0,2)}_M u^a, \nabla^{(0,2)}_M u^b) = (\alpha^m_{rs}, du^a)(\alpha^m_{cd}, du^b)g^{rc}g^{sd}$$

7.9 Barndorff-Nielsen's p^* formula

For a general statistical manifold P a good approximation to the measure $M_{N*}(p\mu)$ is given by the p^* formula of Bardorff-Nielsen which is

$$p^*(p) = C\exp(\ell(p, M^{-1}(p))\mu_g$$

where μ_g is the density on P defined by the Fisher-information metric and the constant C is determined by making p^* a probability measure on P.

This formula obviously requires that the mle be a bijection. This, of course, may not be the case. If it isn't we need to introduce an ancillary statistic

$$a\colon \Omega^N \to \mathbf{R}^{m(N)}$$

with the property that if we restrict attention to a fibre of a then the mle is a bijection. We refer the reader to Barndorff-Nielsen (1988) and references therein for details.

The p^* formula is, in general, a remarkably good approximation to the measure $M_{N*}(p\mu)$. In particular for 'most' transformation models $p^* = M_{N*}(p\mu)$. Again we refer the reader to Barndorff-Nielsen (1988) and references therein for details.

7.10 Remarks for Chapter 7

Remark 7.1 An important application of differential geometry to statistics we haven't considered in elucidating the structure of the 'Bartlett adjustment'. We refer the interested reader to the literature Barndorff-Nielsen and Cox (1984), McCullagh and Cox (1986), Barndorff-Nielsen (1986b) and references therein. An interesting geometric approach appears in Vos (1989).

7.11 Exercises for Chapter 7

Exercise 7.1 Calculate the mle, if it exists, in the von Mises–Fisher model.

Exercise 7.2 Calculate $E(f^* h^*)$ in section 7.8.

CHAPTER 8

Bundles and tensors

8.1 Introduction

So far we have managed to avoid mentioning two of differential geometry's most useful concepts, vector bundles and principal bundles. Vector and principal bundles are generalisations of the tangent bundle bundle of a manifold. We want to discuss them now partly for the sake of completeness but more because they enable us to give a definition of tensor which is readily generalised in our final chapter to higher-order constructions such as strings.

8.2 Tangent bundles and vector bundles

For a smooth manifold P we can also make TP, the tangent bundle to P, into a smooth manifold in a natural way. More precisely, there are natural local co-ordinate systems for TP and they are all smoothly related. If $(\varphi^1, \ldots, \varphi^n)$ is a local co-ordinate system on $U \subset P$ and $p \in U$, then we attach to any tangent vector $v \in T_pP$ the list of $2n$ numbers

$$v \mapsto (\varphi^1(p), \ldots, \varphi^n(p), d\varphi^1(v), \ldots, d\varphi^n(v))$$

This defines a local co-ordinate system on TU, the set $\pi^{-1}(U)$ of all tangent vectors located at points of U. In short, we use the co-ordinates of position and the rates of change of co-ordinate functions to parametrise the set of tangent vectors. If $(\psi^1 \circ \pi, \ldots, \psi^n \circ \pi, d\psi^1, \ldots, d\psi^n)$ is another such co-ordinate system then, since P is smooth, each of $\psi^1 \circ \pi, \ldots, \psi^n \circ \pi$ is a smooth function relative to $(\varphi^1 \circ \pi, \ldots, \varphi^n \circ \pi)$ and so *a fortiori* relative to

$(\varphi^1 \circ \pi, \ldots, \varphi^n \circ \pi, d\varphi^1, \ldots, d\varphi^n)$. Indeed, each $\psi^i \circ \pi$ is independent of $d\varphi^1, \ldots, d\varphi^n$ and depends smoothly on $(\varphi^1 \circ \pi, \ldots, \varphi^n \circ \pi)$. On the other hand we have

$$d\psi^i(v) = \frac{\partial \psi^i}{\partial \varphi^j}(p) d\varphi^j(v) \quad p = \pi(v)$$

from which one can easily deduce that $d\psi^i$ is a smooth function of $(\varphi^1 \circ \pi, \ldots, \varphi^n \circ \pi,\ d\varphi^1, \ldots, d\varphi^n)$ since the functions

$$\frac{\partial \psi^i}{\partial \varphi^j}$$

are smooth on P.

There is another very important way to look at the co-ordinates of tangent vectors. By the values of the location map $TP \xrightarrow{\pi} P$ we divide TP into a family of vector spaces $T_pP = \pi^{-1}\{p\}$. We often call the sets $\pi^{-1}\{p\}$ the fibres of the map π. Given a local co-ordinate system $(\varphi^1, \ldots, \varphi^n)$ we can produce a basis

$$\frac{\partial}{\partial \varphi^1}(p), \ldots, \frac{\partial}{\partial \varphi^n}(p)$$

for each of these vector spaces T_pP by the condition

$$d\varphi^i \left(\frac{\partial}{\partial \varphi^j}(p) \right) = \begin{cases} 1 & i = j \\ 0 & i \neq j \end{cases}$$

We have described before how one then obtains

$$v = d\varphi^i(v) \frac{\partial}{\partial \varphi^i}(p)$$

Hence the co-ordinates of v in TP appear as the co-ordinates $\varphi^1(p), \ldots, \varphi^n(p)$ of its position and the coefficients of its expansion in terms of the basis

$$\frac{\partial}{\partial \varphi^1}(p), \ldots, \frac{\partial}{\partial \varphi^n}(p)$$

of T_pP.

There are many examples which invite an abstraction of this point of view. We are given a set E and a location map $E \xrightarrow{\pi} P$ so that we can think of each element $e \in E$ as having a location $\pi(e)$ in P. Furthermore the set $\pi^{-1}(p)$, which is usually called the *fibre* at p and denoted E_p, is a vector space. A map in the other direction to π, i.e. $P \xrightarrow{s} E$ is called a *section* if, for each $p \in P$, $s(p) \in E_p$. It is the same thing to say

$$\pi \circ s = Id \quad \text{or} \quad \pi(s(p)) = p \quad \forall p \in P$$

since this reads 'position of $s(p)$ is p'. A list of sections $s^1, \ldots, s^m$ is called a *frame* for $E \xrightarrow{\pi} P$, if for each $p \in P$, the values $s^1(p), \ldots, s^n(p)$ are a basis for E_p. In this terminology

$$\frac{\partial}{\partial \varphi^1}, \ldots, \frac{\partial}{\partial \varphi^n}$$

are sections of $TU \xrightarrow{\pi} U$ where $U \subset P$ is the domain of the local co-ordinate system $\varphi^1, \ldots, \varphi^n$. A section $X : U \to TU$ assigns to each point $p \in U$ a tangent vector $X(p) \in T_pP$ located at p. Sections of $TU \xrightarrow{\pi} U$ are usually called *vector fields*. Because each

$$\frac{\partial}{\partial \varphi^i}$$

is a section defined on only a subset of P is it often called a *local section* of $TP \xrightarrow{\pi} P$. From our earlier discussion it is clear that

$$\frac{\partial}{\partial \varphi^1}, \ldots, \frac{\partial}{\partial \varphi^n}$$

is a local frame for $TP \to P$. In the general case $E \xrightarrow{\pi} P$, given a local co-ordinate system $\varphi^1, \ldots, \varphi^n$ on $U \subset P$ and a frame $s^1, \ldots, s^m$ for $E_U \xrightarrow{\pi} U$ where $E_U = \pi^{-1}(U)$, we can expand any $e \in E_U$ as

$$e = \lambda_1(e)s^1(p) + \ldots + \lambda_m(e)s^m(p) \quad p = \pi(e)$$

and attach to e the list of $m + n$ numbers

$$e \mapsto (\varphi^1(p), \ldots, \varphi^n(p), \lambda_1(e), \ldots, \lambda_m(e))$$

We thereby obtain a co-ordinate system on E_U.

As an important example of such a situation we consider the set T^*P which is the union of the duals of the tangent spaces

$$T^*P = \bigcup_{p\in P} (T_pP)^*$$

An element θ of T^*P is thus a linear function on some tangent space T_pP, and we call p the location of θ. Hence T^*P has a natural location map $T^*P \xrightarrow{\pi} P$ and the set $\pi^{-1}(p)$ is precisely $(T_pP)^*$ which is a vector space. It is also denoted T_p^*P. Differential 1-forms are most economically described as sections of $T^*P \xrightarrow{\pi} P$. Recall that a differential 1-form ω is a function on TP whose restriction ω_p to each tangent space T_pP is linear. We should therefore think of ω_p as belonging to T_p^*P, and ω as a map $P \to T^*P$ whose value at p is $\omega_p \in T_p^*P$. For each smooth function f we have a section df, and if $(\varphi^1, \ldots, \varphi^n)$ is a local co-ordinate system on $U \subset P$ then $d\varphi^1, \ldots, d\varphi^n$ is a frame for T^*U. We thereby obtain a local co-ordinate system on T^*U. It is the one which assigns to each $\theta \in T^*U$ the co-ordinates of its location p, along with the co-efficients of its expansion as a linear combination of $d\varphi_p^1, \ldots, d\varphi_p^n$.

The special local co-ordinate systems on E_U defined by a local frame and local co-ordinate system on U are called *local trivialisations* in the literature – the reasons need not concern us for now. When E is a smooth manifold whose maximal atlas contains a sub-atlas of local co-ordinate systems which are local trivialisations, we call $E \xrightarrow{\pi} P$ a *vector bundle.* To emphasise this fact for the set of all tangent vectors, TP is called the *tangent bundle* of P. Of course, we can start with a family of local trivialisations on E which are a smooth atlas and extend it to a maximal atlas. In the case of T^*P we can show that the natural local trivialisations are smoothly related. Thus we generate a maximal atlas making T^*P a smooth manifold and $T^*P \to P$ a vector bundle called the cotangent bundle.

In order to establish that two local trivialisations are smoothly related we have the following important fact. Suppose that one is defined by a local co-ordinate system $(\varphi^1, \ldots, \varphi^n)$ on U and a frame $s^1, \ldots, s^m$ on U. Let the other be defined by $(\psi^1, \ldots, \psi^n)$ and $t^1, \ldots, t^n$ similarly. We suppose that $(\varphi^1, \ldots, \varphi^n)$ and $(\psi^1, \ldots, \psi^n)$ are smoothly related. At each $p \in U$ we have two bases $s^1(p), \ldots,$

$s^m(p)$ and $t^1(p),\ldots,t^m(p)$ for E_p. We therefore have a matrix $f_{ij}(p)$ defined by

$$t^i(p) = \sum_{j=1}^{n} f^i_j(p)s^j(p)$$

This defines a matrix of functions f^i_j on U, and we say that the frames are *smoothly related* if the functions f^i_j are smooth. The important result, which is easily proved, is that two local trivialisations are smoothly related if and only if the local co-ordinate systems and frames which define them are smoothly related.

The simple argument leading to this result involves ideas which are important in their own right. Given a map $F : P \to N$ between two smooth manifolds we say that it is smooth if for each local co-ordinate system $\psi^1,\ldots,\psi^n$ the functions $\psi^1 \circ F,\ldots,\psi^n \circ F$ are smooth. In particular, given a vector bundle $E \xrightarrow{\pi} P$ we can consider its set of smooth sections, and these are usually denoted $\Gamma(E)$. By the definitions involved the sections $s^1,\ldots,s^m$ defining a local trivialisation in the atlas of E are smooth sections. Hence one may say that the vector bundle structure of TP is so designed that for each local co-ordinate system $(\varphi^1,\ldots,\varphi^n)$, the sections

$$\frac{\partial}{\partial\varphi^1},\ldots,\frac{\partial}{\partial\varphi^n}$$

are smooth. Similarly the vector bundle structure of T^*P is so designed that

$$d\varphi^1,\ldots,d\varphi^n$$

are smooth. Given any frame $s^1,\ldots,s^m$ over a set $U \subset P$ for a vector bundle E, we can expand each section s over U pointwise. For each $p \in U$ this produces coefficients $f_1(p),\ldots,f_m(p)$ such that

$$s(p) = f_1(p)s^1(p) + \ldots + f_m(p)s^m(p) \quad p \in U$$

The functions $f_1,\ldots,f_m$ so defined are called the coefficient functions of s for its expansion in terms of $s^1,\ldots,s^m$. It is an important fact that if $s^1,\ldots,s^m$ is a frame then s is smooth if and only if its coefficient functions are smooth.

This fact allows an easy test for the smoothness of vector fields or differential 1-forms. Given any differential 1-form ω we choose a local co-ordinate system $(\varphi^1, \ldots, \varphi^n)$ on a set U and expand ω as

$$\omega = f_1 d\varphi^1 + \ldots + f_n d\varphi^n$$

Then ω is smooth on U if and only if $f_1, \ldots, f_n$ are smooth. Similarly, for a vector field X we expand

$$X = f^1 \frac{\partial}{\partial \varphi^1} + \ldots + f^n \frac{\partial}{\partial \varphi^n}$$

Then X is smooth on U if and only if $f^1, \ldots, f^n$ are. Returning to a general vector bundle $E \xrightarrow{\pi} P$, these arguments also show that the sum of two smooth sections $s_1, s_2 \in \Gamma(E)$ defined pointwise by

$$(s_1 + s_2)(p) := s_1(p) + s_2(p) \in E_p$$

is also smooth. Furthermore the product of a smooth section $s \in \Gamma(E)$ with a smooth function f, also defined pointwise by

$$(fs)(p) := f(p)s(p) \in E_p$$

is also smooth. Hence the set $\Gamma(E)$ of smooth sections of any vector bundle admits linear operations, that is, an addition and scalar multiplication, where the scalars are smooth functions.

Example 8.2.1 *The tangent bundle.* Consider the tangent bundle $T\mathcal{P}$ to the space of all probability measures. This is a bundle with infinite-dimensional fibres and we should really be careful about its structure as a manifold but as usual we shall be careless and say nothing. We can think of this tangent bundle in two different ways. On the one hand, because $\mathcal{P}$ is the space of measures up to scale we can identify its tangent space at any point with, $R_\Omega/\mathbf{R}.1$, the space of randon variables up to addition of constants. It follows that the tangent bundle to $\mathcal{P}$ is isomorphic to a trivial bundle

$$T\mathcal{P} \simeq R_\Omega/\mathbf{R}.1 \times \mathcal{P}$$

On the other hand $\mathcal{P}$ is a submanifold of $\mathcal{M}$ so that each $T_p\mathcal{P}$ is a subspace of $T_p\mathcal{M} = R_\Omega$, the subspace of random variables of zero mean. That is

$$T_p\mathcal{P} = \{(p, f) \mid E_p(f) = 0\} \subset \mathcal{P} \times R_\Omega$$

We can identify the fibres of these bundles as follows. If f is a random variable with $E_p(f) = 0$ then we map it to the equivalence class $f + \mathbf{R}.1$. On the other hand if $f + \mathbf{R}.1$ is an equivalence class we map it to $f - E_p(f).1$ and this is clearly well defined.

We can now restrict this bundle to a bundle over P. If we assume further that we can choose the space of random variables R_Ω so that they are square integrable for all p in $\mathcal{P}$ then each fibre can be completed to a Hilbert space with the usual expectation inner product

$$\langle f, g \rangle_p = E_p(fg)$$

We denote the resulting bundle of Hilbert spaces over P by $\mathcal{H}$. This is Amari's Hilbert bundle (see for instance Amari et. al. (1987)).

Example 8.2.2 *The normal bundle.* If $P \subset \mathcal{P}$ is a statistical manifold and we assume that $T_pP \subset \mathcal{H}_p$ for all p then we can define its normal N_p. The collection N of all these vector spaces defines a vector bundle over P called the normal bundle to P in $\mathcal{P}$. The second fundamental form is a bilinear map

$$\alpha : TP \times TP \to N$$

from the tangent bundle to the normal bundle.

8.3 New vector bundles from old

A number of algebraic constructions that can be applied to vector spaces can also be applied to vector bundles. It will be useful for us to point out a number of these.

If E and F are vector bundles over P we can define a vector bundle $E \oplus F$ whose fibre at p is $E_p \oplus F_p$, the direct sum of the fibres of E and F at p. We call $E \oplus F$ the direct sum of E and F. Notice that we must not confuse this with $E \times F$, the Cartesian product of E and F as manifolds. Indeed $E \oplus F$ is a submanifold of $E \times F$ defined by

$$E \oplus F = \{(e, f) \mid \pi(e) = \pi(f)\} \subset E \times F$$

To prove that this is a vector bundle we just note that if $(e^1, \ldots, e^n)$ and $(f^1, \ldots, f^m)$ are local frames for E and F then

$$(e^1, 0), (e^2, 0), \ldots, (e^n, 0), (0, f^1), \ldots, (0, f^n)$$

is a local frame for $E \oplus F$ and we can readily check that these local frames all fit together smoothly.

Similarly if E and F are vector bundles over P we can define a vector bundle $E \otimes F$ whose fibre at p is $E_p \otimes F_p$, the tensor product of the fibres of E and F at p. We call $E \otimes F$ the tensor product of E and F. Recall that if V and W are two vector spaces then we can define a vector space $V \otimes W$, the tensor product of V and W Koboyashi and Nomizu (1963). The elements of $V \otimes W$ are linear combinations of the elements $v \otimes w$ obtained by forming the tensor product of an element of V and an element of W. This tensor product is bilinear in the sense that

$$(av + bv') \otimes w = a(v \otimes w) + b(v' \otimes w)$$

and

$$v \otimes (aw + bw') = a(v \otimes w) + b(v \otimes w')$$

To prove that $E \otimes F$ is a vector bundle we just note that if e^a and f^b are local frames for E and F then $u^{ab} = (e^a \otimes f^b)$ is a local frame for $E \oplus F$ and we can readily check that these local frames all fit together smoothly.

Lastly if E is a vector bundle we can define the dual bundle by taking the dual of every fibre. That is $(E^*)_p = E^*_p$.

8.4 Connections on a vector bundle

We have already introduced the notion of a connection on the tangent bundle. It is easy to generalise this now to the case of a vector bundle E. Denote by $\Gamma(E)$ the space of all sections of E. Then a connection is a linear map

$$\nabla : \Gamma(E) \to \Gamma(T^* \otimes E)$$

which satisfies the Leibniz rule:

$$\nabla(f\psi) = df \otimes \psi + f\nabla\psi$$

A common method of constructing connections is as follows. Assume that the bundle E is a sub-bundle of a trivial bundle $W \times P$ for some vector space W and that, moreover, at each p there is an inner product $\langle\ ,\ \rangle_p$ on W. In other words each fibre E_p is a subspace of W. Then if ψ is a section of E we can think of it as map into W and differentiate it in the usual way. If X is a tangent vector this derivative is just $d_X\psi$. In general, $d_X\psi$ will not be in E_p only in W. However the inner product defines an orthogonal projection $\pi_p: W \to E_p$ and we can define

$$\nabla_X(\psi) = \pi_p(d_X\psi)$$

That this obeys the Leibniz rule is a consequence of the fact that d does.

8.5 Frame bundles and principal bundles

A choice of co-ordinates $\theta^1, \ldots, \theta^n$ on a manifold defines a basis, or frame,

$$\frac{\partial}{\partial\theta^1}(p), \ldots, \frac{\partial}{\partial\theta^n}(p)$$

for each tangent space T_pP. It is is extremely useful to introduce the space of all frames for the tangent space.

Before doing this consider an n-dimensional vector space V. Denote by $F(V)$ the set of all frames for V. A useful way of understanding what $F(V)$ looks like is to note that $GL(n)$ acts freely and transitively on $F(V)$. By convention we are going to make this an action on the right. If $X = X^i_j$ is an element of $GL(n)$ and $v = \{v_1, \ldots, v_n\}$ is a basis of V then we define vX by

$$(vX)_i = \sum_{j=1}^{n} v_j X^j_i$$

The reader should check that this is a right action, that is that $(vX)Y = v(XY)$. If w and v are two frames then $w = vX$ where X is defined uniquely by

$$w_i = \sum_{j=1}^{n} v_j X^j_i$$

So $F(V)$ as a manifold is diffeomorphic to $GL(n)$.

There is another way of defining $F(V)$ which is more useful for the generalisations we shall want in the next chapter. This approach defines $F(V)$ to be the set of all linear isomorphisms $\phi: V \to \mathbf{R}^n$. Then if we note that $GL(n)$ is the set of all linear isomorphisms $\mathbf{R}^n \to \mathbf{R}^n$ we see that the action we have defined is

$$\phi X = \phi \circ X$$

where $\circ$ is composition.

Notice that if we pick a point ϕ of $F(V)$ we immediately have a canonical diffeomorphism

$$\begin{array}{ccc} GL(V) & \simeq & F(V) \\ X & \mapsto & \phi X \end{array}$$

At a point p in a manifold P define $F_p(P)$ to be $F(T_pP)$. That is the set of all linear isomorphisms from $\mathbf{R}^n$ to T_pP. Then define the frame bundle $F(P)$ to be the union over P of all the $F_p(P)$. Notice that every element of $F(P)$ is a frame at some point of P and this defines a map $\pi: F(P) \to P$. So we have the diagram

$$\begin{array}{cc} & F(P) \\ \pi & \downarrow \\ & P \end{array}$$

This diagram illustrates the conventional way to think of the frame bundle. We think of it as a family of vertical fibres (the $F_p(P)$) which sit above the horizontal manifold P.

To summarise $F(P)$ is a manifold on which $GL(n)$ acts with an onto map $\pi: F(P) \to P$ such that the fibres of π are precisely the orbits of G. We want to abstract the properties of the frame bundle as we abstracted the properties of the tangent bundle to define vector bundles, and what we have just said is not quite enough. The frame bundle has one more important property usually called *local triviality*. Let us explain what this means. A *section* of the frame bundle is a map $s: P \to F(P)$ which is like a section of a vector bundle in as much as its value $s(p)$ at any point p is in the space $F_p(P)$, that is the fibre of $F(P)$ above p. Unlike the case of vector bundles frame bundles may not have *any* sections. As an example consider the 2-sphere. A section of the frame bundle of the 2-sphere is a choice of a frame in every tangent space. In particular this means a choice of a non-vanishing vector field but it is a result in topology that the sphere has no non-vanishing vector fields. However we know that we can chose local co-ordinates for P and then on each co-ordinate neighbourhood we have the frames

$$\frac{\partial}{\partial\theta^1}, \ldots, \frac{\partial}{\partial\theta^n}$$

which are sections over that co-ordinate neighbourhood. We say that the frame bundle admits local sections. If the frame bundle has a global section s we also say that it is *trivial* as we can define a diffeomorphism

$$\chi: P \times GL(n) \to F(P)$$

which satisfies $\chi(m, X) = \chi(m)X$ by

$$\chi(m, X) = s(m)X$$

Conversely given any such χ we can define s, a global section, by $s(m) = \chi(m, 1)$.

The abstraction of the frame bundle is called a principal bundle. This consists of three manifolds $F(P, G)$ where G is a Lie group acting freely on the right of P, there is an onto map $\pi: F \to P$ whose fibres are precisely the orbits of G and π admits local sections. The simplest such principal bundle is $P \times G(P, G)$ with the right action $(p, h)g = (p, gh)$. We call a principal bundle $F(P, G)$ trivial if it is isomorphic to the trivial bundle. That is there is a diffeomorphism $\chi: P \times G \to F$ with $\chi(m, g) = \chi(m)g$ and $\pi \circ \chi(p, g) = p$. That is χ preserves the G action in both bundles and maps the fibre over p in one bundle to the fibre over G in the other. We have just seen that a principal bundle is trivial if and only if it admits a global section. This finally explains why we say that $F(P)$ is locally trivial.

The simplest example of a principal bundle which is not trivial is to construct the Moebius band. Draw a circle in the middle of the band and call this P. Now call the edge of the band F and let the group G be the group $Z_2 = \{1, -1\}$ with two elements which acts by flipping a point on one edge to the opposite edge. Notice that F is a circle, not, two circles as it would be for the trivial Z_2 bundle $S^1 \times Z_2$ over the circle S^1.

The manifold structure of F is defined using local triviality. For any co-ordinate chart on $U \subset P$ we have a bijection $F(U) \to$

$U \times GL(n)$. This makes $F(U)$ a manifold and it just has to be checked, in a similar way to the case of vector bundles, that these various ways of making open subsets of $F(P)$ a manifold all fit together to make $F(P)$ a manifold.

All that we have said here generalises immediately to show that is E is a vector bundle whose fibre has dimension r then

$$F(E) = \cup_{p \in P} F(E_p)$$

is a principal bundle with group $GL(r)$.

Example 8.5.1 *The infinite frame bundle.* An example that is useful in statistics is the infinite frame bundle. Let p be a point of our manifold. Let ϕ be a co-ordinate system about p. Then we can expand any function f about p in a Taylor series relative to the co-ordinates ϕ and we obtain a string of numbers which are the coefficients of the Taylor series, that is, the higher derivatives of f. Let us call two co-ordinate systems (infinite jet) equivalent if any function expanded in the two co-ordinate systems has the same Taylor coefficients. We shall call the equivalence class of such a co-ordinate system an infinite frame at p. The reason for this choice of name is that we can also define r frames by requiring that only coefficients up to order r are the same. If we take $r = 1$ we see that co-ordinates ϕ and θ are equivalent if and only if

$$\frac{\partial f}{\partial \phi^i} = \frac{\partial f}{\partial \theta^i}$$

for all i. This is true if and only if

$$\frac{\partial}{\partial \phi^i} = \frac{\partial}{\partial \theta^i}$$

for all i. Hence an equivalence class is equivalent to a frame for T_pP.

Just as the 1-frames form a principal bundle for $GL(n)$ the infinite frames form a principal bundle for a group $\mathcal{G}_\infty(n)$ which we shall call the ***infinite phylon group***. This group is constructed as follows. Consider the group of all local diffeomorphisms of $\mathbf{R}^n$ which fix the origin. If X is such a diffeomorphism then it acts on the set of all frames at p by

$$\phi X = X^{-1} \circ \phi$$

The inverse here is to make the group action a right action. This action is not free because any diffeomorphism which differs from the identity by a map from $\mathbf{R}^n$ to $\mathbf{R}^n$ all of whose derivatives are zero acts trivially. So we form the quotient group obtained by saying that two diffeomorphisms are (infinite jet) equivalent if they have the same Taylor series about zero. The group $\mathcal{G}_\infty(n)$ is the group of all these equivalence classes. Equivalently it is the group of all formal power series

$$f = f^j_{i_1} x^{i_1} + \frac{1}{2!} f^j_{i_1 i_2} x^{i_1 i_2} + \ldots \qquad (8.5.1)$$

where f^j_i is an invertible matrix and the multiplication is substitution of one power series into the other.

If $\mathcal{F}^\infty_p(P)$ is the space of all infinite frames at p then

$$\mathcal{F}^\infty(P) = \bigcup_{p \in P} \mathcal{F}^\infty_p(P)$$

is a principal bundle for $\mathcal{G}_\infty(n)$. If we take an open set U in P and some co-ordinates θ on U then $\theta_p(q) = \theta(p) - \theta(q)$ defines a co-ordinate system about p and therefore a section of $\mathcal{F}^\infty(P)$ over U. If we specified carefully the differentiable structure of $\mathcal{F}^\infty(P)$ then these would be local sections.

If we replace ∞ by r then there is an obvious definition for $\mathcal{G}_r(n)$. The bundle $\mathcal{F}^r(P)$ is a principal bundle for $\mathcal{G}_r(n)$. In this case it is finite dimensional and we can give it the differentiable structure that makes the local sections about local trivialisations.

In the case that $r = 1$ then $\mathcal{G}_1(n)$ is just the group $GL(n)$ and we recover the usual frame bundle.

8.6 Associated bundles

We have seen in the previous section how to construct principal bundle of frames $F(E)$ of a vector bundle E. In this section we want to show how to reverse this procedure. Consider the simplest case where we have a vector space V and the manifold $F(V)$. How do we get back from $F(V)$ to V ? First we note that $GL(r)$ acts naturally on $\mathbf{R}^r$ and also on $F(V)$. We can therefore make it act on the right of the product $F(V) \times \mathbf{R}^r$ by $(f, v)X = (fX, X^{-1}v)$ and form the space of orbits which we denote by

$$F(V) \times_{GL(r)} \mathbf{R}^r$$

We shall use the notation $[f, v]$ for the orbit containing (f, v) that is

$$[f, v] = \{(fX, X^{-1}v) | X \in GL(r)\}$$

In particular this means that for any X in $GL(r)$ we have $[f, v] = [fX, X^{-1}v]$.

Let us denote by $F(V) \times_{GL(r)} \mathbf{R}^r$ this space of orbits. To understand what it looks like choose some f_0 in $F(V)$. Then for every orbit $[f, v]$ such that $\pi(f) = p$ we have $f_O X = f$ for some X in $GL(r)$. It follows that the map

$$\begin{array}{ccc} \mathbf{R}^r & \to & F(V) \times_{GL(r)} \mathbf{R}^r \\ w & \mapsto & [f_0, w] \end{array}$$

is a bijection. In fact it is not difficult to show that $F(V) \times_{GL(r)} \mathbf{R}^r$ is a vector spaces and that this map is a linear isomorphism of vector spaces. Take two orbits $[f_0, v]$ and $[f_0, w]$ and define

$$\lambda[f_0, v] + \mu[f_0, w] = [f_0, \lambda v + \mu w]$$

for any λ and μ in $\mathbf{R}$. To see that this definition is independent of the choice of f_0 replace f_0 by $f_0 X$ for X in $GL(r)$. Then we have

$$\begin{aligned} \lambda[f_0, v] + \mu[f_0, w] &= \lambda[f_0 X, X^{-1}v] + \mu[f_0 X, X^{-1}w] \\ &= [f_0 X, \lambda X^{-1}v + \mu X^{-1}w] \\ &= [f_0 X, X^{-1}(\lambda v + \mu w)] \\ &= [f_0, \lambda v + \mu w] \end{aligned}$$

The important point here is that X acts linearly on $\mathbf{R}^r$ so we have that

$$\lambda X^{-1}v + \mu X^{-1}w = X^{-1}(\lambda v + \mu w)$$

Now consider an element $[f, v]$ in $F(V) \times_{GL(r)} \mathbf{R}^r$ and note that by definition f is an isomorphism $f: \mathbf{R}^n \to V$ where $\pi(f) = p$. We then define

$$\begin{array}{ccc} F(V) \times_{GL(r)} \mathbf{R}^r & \to & V \\ [f, v] & \mapsto & f(v) \end{array}$$

It is straightforward to check that this is independent of the choice of (f, v) in the orbit and that is linear. It is therefore an isomorphism.

It is perhaps easier to understand what is going on here if we think of $F(V)$ as the set of all bases for V. Then every vector v has a list of components $v_1(f), \ldots, v_r(f)$ in any frame f. If we know what these components are in some frame then of course we know what v is, namely

$$v = \sum_i v_i(f) f_i$$

If we know that the components of all vectors in all frames then that is the space $F(V) \times V\mathbf{R}^r$ and to recover V we have to identify pairs (f, v) and (f', v') for which

$$\sum_i v_i'(f') f_i' = v = \sum_i v_i(f) f_i$$

but this happens only when $f' = fX$ and $v' = X^{-1}v$.

Now if we have a vector bundle E instead of a vector space V then we just have to do this construction 'fibre by fibre'. Given the principal bundle $F(E)$ then $GL(r)$ acts on the space $F(E) \times \mathbf{R}^r$ by $(f, v)X = (fX, X^{-1}v)$. Denote by $F(E) \times_{GL(r)} \mathbf{R}^r$ the space of orbits of this action and again use the notation $[f, v]$ for the orbit containing (f, v) that is

$$[f, v] = \{(fX, X^{-1}v) | X \in GL(r)\}$$

We have to show first that $F(E) \times_{GL(r)} \mathbf{R}^r$ is a vector bundle over P. The projection map is

$$\begin{array}{ccc} F(E) \times_{GL(r)} \mathbf{R}^r & \to & P \\ [f, v] & \mapsto & \pi(f) \end{array}$$

where $\pi : F(E) \to P$ is the projection. We have to be rather careful with this definition as given an orbit $[f, v]$ there is no distinguished element in it so it is not clear that what we have done is well defined. In this case we note that if $[f, v] = [f', v']$ then there is an X in $GL(r)$ such that $f' = fX$ and $v' = X^{-1}v$. Then is we apply π to either f or f' we get the same result. Recall that $F_p(E) = F(E_p)$ so that fibre over p is just $F(E_p) \times \mathbf{R}^r$. It follows from the vector space case that this is a vector space with scalar multiplication and addition defined by

$$\lambda[f_0, v] + \mu[f_0, w] = [f_0, \lambda v + \mu w]$$

for any λ and μ in $\mathbf{R}$.

To finally conclude that $F(E) \times_{GL(r)} \mathbf{R}^r$ is a vector bundle we have show that it admits local frames. It is enough to note that if $e_1, \ldots, e_r$ is a basis of $\mathbf{R}^r$ and s is a section of F then $[s, e_1], \ldots, [s, e_r]$ is a collection of sections of $F(E)$. Notice that for any p at which s is defined the map $w \mapsto [s(p), w]$ is an isomorphism so that the vectors $[s(p), e_1], \ldots, [s(p), e_r]$ are a basis of the fibre of $F(E)$ over p. Now it follows from the fact that $F(E)$ has local sections that $F(E) \times_{GL(r)} \mathbf{R}^r$ has local frames. Notice that we have not said how to make $F(E) \times_{GL(r)} \mathbf{R}^r$ a manifold. The simplest way to do this is to use the local frames as in the case of vector bundles in section 8.2. We leave the details as an exercise for the reader.

To show that the vector bundles $F(E) \times_{GL(r)} \mathbf{R}^r$ and the V are isomorphic we just repeat the construction we used in the vector space case. We consider an element $[f, v]$ in $F(E) \times_{GL(r)} \mathbf{R}^r$ and note that by definition f is an isomorphism $f: \mathbf{R}^n \to E_p$ where $\pi(f) = p$. We then define

$$\begin{array}{ccc} F(E) \times_{GL(r)} \mathbf{R}^r & \to & E \\ {[f, v]} & \mapsto & f(v) \end{array}$$

It is straightforward to check that this is an isomorphism of vector bundles.

The construction of vector bundles we have given here can be generalised as follows. Let $Z(P, G)$ be any principal bundle and choose a representation $\phi: G \to GL(V)$ for some vector space V. Then we can make G act on the right of $Z \times V$ by $(z, v)g = (zg, \phi(g^{-1})v)$. It is a straightforward generalisation of what we have already done to see that the space of orbits $Z \times_G V$ is a vector bundle. We call the bundle $Z \times_G V$ the vector bundle associated to $Z(P, G)$ by the representation ϕ.

8.7 Tensors

We have already seen examples of tensors such as the Fisher information metric. Let us give a general definition which will motivate the next chapter. Recall the definition of the tensor product of two vector spaces (section 8.3). If V is a vector space denote by $T^{(r,s)}(V)$ the vector space

$$T^{(r,s)}(V) = \otimes^r(V) \otimes \otimes^s(V^*)$$

where

$$\otimes^r(V) = \underbrace{V \otimes \cdots \otimes V}_{r \text{ times}}$$

If P is a manifold we denote by $T^{(r,s)}(P)$ the vector bundle whose fibre at p is $T^{(r,s)}(T_pP)$. The elements of $T^{(r,s)}(\mathbf{R})n$ are multi-indexed objects $T = T^{i_1 \dots i_r}_{j_1 \dots j_s}$ where all indices run from 1 to n. We can make $GL(n)$ act on such objects by defining the action of $X = X^i_j$ by

$$(XT)^{i_1 \dots i_r}_{j_1 \dots j_s} = T^{k_1 \dots k_r}_{l_1 \dots l_s} X^{i_1}_{k_1} \dots X^{i_r}_{k_r} \hat{X}^{l_1}_{j_1} \dots \hat{X}^{l_s}_{j_s}$$

where $\hat{X}$ is the inverse transpose matrix of X. This defines a representation $\tau^{(r,s)}$ of $GL(n)$ on $T^{(r,s)}(\mathbf{R}^n)$. We can therefore apply the associated bundle construction to define a vector bundle

$$F(P) \times_{GL(n)} T^{(r,s)}(\mathbf{R}^n)$$

We leave it as an exercise for the reader to generalise the proof in section 8.6 for the tangent bundle to show that this bundle is the vector bundle $T^{(r,s)}(P)$.

We should really be careful to distinguish two things here. There are the elements of the fibre $T^{(r,s)}(P)_p$ and there are sections of the bundle $T^{(r,s)}(P)$. When it is important to make the distinction we shall call the former tensors and the latter tensor fields. Often, however, we shall follow tradition and call them both just tensors.

Classically tensor fields were often specified locally. To see what this means we choose local co-ordinates θ with domain U_θ. Then at each point of U_θ the tensors

$$\frac{\partial}{\partial \theta^{i_1}} \otimes \cdots \otimes \frac{\partial}{\partial \theta^{i_r}} \otimes d\theta^{j_1} \otimes \cdots \otimes d\theta^{j_s}$$

define a basis of $T^{(r,s)}(P)_p$. If ξ is a tensor field on P, that is a section of the bundle $T^{(r,s)}(P)$, then we can expand it in this basis as

$$\xi = \sum \xi^{i_1,\ldots,i_r}_{j_1,\ldots,j_s} \frac{\partial}{\partial\theta^{i_1}} \otimes \cdots \otimes \frac{\partial}{\partial\theta^{i_r}} \otimes d\theta^{j_1} \otimes \cdots \otimes d\theta^{j_s}$$

We call the collection of functions

$$\xi^{i_1,\ldots,i_r}_{j_1,\ldots,j_s}$$

the components of ξ with respect to the co-ordinates θ. We can also think of these components as a map

$$\xi(\theta)\colon U_\theta \to T^{(r,s)}(\mathbf{R}^n)$$

Classically a tensor was specified as a collection of such components, one for each co-ordinate system. When two co-ordinate systems overlap the co-ordinate expressions have to be related. It is linear algebra to check that the relationship is

$$\xi(\chi)^{i_1\ldots i_r}_{j_1\ldots j_s} = \xi(\theta)^{k_1\ldots k_r}_{l_1\ldots l_s} \frac{\partial\chi^{i_1}}{\partial\theta^{k_1}} \cdots \frac{\partial\chi^{i_r}}{\partial\theta^{k_r}} \frac{\partial\theta^{l_1}}{\partial\chi^{k_1}} \cdots \frac{\partial\theta^{j_1}}{\partial\chi^{j_s}} \qquad (8.7.1)$$

Classically it would have been said that the tensor transforms by this rule. It is horrible formulae like this that have given tensor analysis a bad name. It is helpful to remember that this is nothing but the linear algebra associated with expanding an element ξ of the vector space $T^{(r,s)}(P)_p$ in two different bases. This formulae can be summarised in various more succinct ways such as

$$\xi(\chi) = \tau^{(r,s)}(\frac{\partial\chi^i}{\partial\theta^j})\xi(\theta)$$

where the i and j indices are taked to define a matrix in $GL(n)$.

Given that we can construct all the tensor bundles as tensor products of vector bundles the reader is justified in asking why we would bother with the theory of frame bundles. One answer is that it enables us to make precise the question 'what are all the posssible tensors in the sense of things transforming according to (8.7.1) ?' Namely we can ask what are all the vector bundles we can associate to the frame bundle. This is equivalent to the question what are all the (finite-dimensional) representations of $GL(n)$. The representation theory of $GL(n)$ is well known Humphreys (1981) but to state the answer we need to first define some concepts. If V is a representation of a group G we say it is irreducible if it has no invariant subspaces, that is subspaces which are fixed by all the elements of $GL(n)$. We say it is decomposable if it can be written as a direct sum of two other representations. Every finite-dimensional representation is completely decomposable, that is can be written as a direct sum of irreducibles so it suffices to classify the irreducibles. The tensor representations are in general not irreducible but amongst their irreducible summands occur all the possible irreducible representations of $GL(n)$. In this sense the answer to this question is that all possible tensors occur as sections of some $T^{(r,s)}(P)$.

Example 8.7.1 *The Fisher information.* The tensor that arises most often in statistical theory is the Fisher information. This is a section of $T^{0,2}(P)$. Note, however, it is not any old section: it is one which is symmetric and satisfies $g(X, X) > 0$ unless $X = 0$, that is it defines a positive definite inner product.

There are many other tensors that can be formed out of the log-likelihood. For instance we have the *skewness tensor*

$$T(X, Y, Z) = E_p(X(\ell)Y(\ell)Z(\ell))$$

which has components

$$E_p(\frac{\partial \ell}{\partial \theta^1 i}\frac{\partial \ell}{\partial \theta^j}\frac{\partial \ell}{\partial \theta^k})$$

This tensor is related to the α-connections of Amari. We have

$$g(X, \nabla_Y^{(\alpha)}(Z)) = g(X, \nabla_Y^{(0)}(Z)) - \frac{\alpha}{2}T(X, Y, Z)$$

The theory of a Riemannian manifold with a tensor of this type has been developed by Lauritzen in Amari et al. (1985).

8.8 Remarks for Chapter 8

Remark 8.1 The classic reference for the theory of principal and vector bundles is Koboyashi and Nomizu (1963). This gives a thorough treatment of connections.

8.9 Exercises for Chapter 8

Exercise 8.1 If E and F are vector bundles prove that $E \otimes F$, $E \oplus F$ and E^* are vector bundles.

Exercise 8.2 Calculate the skewness tensor for exponential families and hence, or directly, for the normal family.

Exercise 8.3 Show that the skewness tensor for the unit normal family is zero.

Exercise 8.4 By comparing the observed α-connections for different values of the parameter α define, by analogy with example 8.7.1, an observed skewness tensor.

Exercise 8.5 Repeat exercise 8.4 for a general yoke.

CHAPTER 9

Higher-order geometry

9.1 Introduction

We have seen that Taylor series play a large role in the theory of asymptotic statistics. In $\mathbf{R}^n$ we know how to define a Taylor expansion and what it means for a function to vanish with order p at a point. We want to extend these notions to a manifold.

9.2 Jets and jet bundles

Often we need to talk about a function defined in a neighbourhood of a point p in a manifold. We shall say 'let f be a function about p' rather than talking precisely about the domain of f and if we are comparing two functions about p we shall say they are 'equal about p' if they agree on smaller domain. The precise way to do this is to say that two functions about p are germ equivalent at p if they agree on some open set about p. This is an equivalence relation and the equivalence classes are called germs. So sometimes when we talk about functions we really mean germs of functions.

We say that real-valued functions f and g about p are the same to order r or that they are r-jet equivalent at p if in some co-ordinates they have the same Taylor series up to and including derivatives of order r. The important thing about this concept is that an application of the chain rule shows that it is independent of the choice of co-ordinates. Another way to say this is that the function $f - g$, in some co-ordinates about p, has all the terms up to and including the terms of order r in its Taylor series equal to zero. A few calculations with the chain rule and low-order Taylor series is the simplest way to convince oneself of these facts. Such experiments will also show that saying that the rth terms in the Taylor expansions are equal *is not* generally independent of co-ordinates. Notice that this idea includes the idea of paths having the same tangent vector. Indeed two paths $\gamma, \rho: (-\epsilon, \epsilon) \to P$ have the same tangent vector if they agree to first order at zero.

We can also apply this idea to maps between manifolds if we express the map as a function of many variables with respect to co-ordinates.

If two functions are r-jet equivalent for all r then we say they are infinitely jet equivalent at p. When we say jet equivalent without reference to r we mean infinitely jet equivalent.

Jet equivalence is an equivalence relation and the equivalence class of a function is called its r-jet. The notion of the jet of a function on a manifold is the nearest thing that one can get to the idea of the Taylor series of a function without choosing additional structure. Of course it is essentially all Taylor series in all co-ordinates. Again we stress that the individual terms in the Taylor series have no invariant meaning without choosing extra structure.

This is not quite true! The zeroth- and first-order terms can be defined invariantly. Two functions f and g are 0-jet equivalent at p if $f(p) = g(p)$ and they are 1-jet equivalent if they have the same value and $d_p f = d_p g$. So the 'first-order part' of the function is $d_p f$.

The set of all r-jets of functions at p is a vector space denoted by $J^r_p(P, \mathbf{R})$ and it is finite-dimensional. It is not hard to see that, just as in the case of the tangent bundle, the union of all these jet spaces

$$J^r(P, \mathbf{R}) = \cup_{p \in P} J^r_p(P, \mathbf{R})$$

is a vector bundle over P.

Example 9.2.1 *The second jet bundle.* Consider the second jet space $J^2(P, \mathbf{R})_p$. If θ are local co-ordinates on an open set U containing p then denote by J_U the set

$$J_U = \bigcup_{p \in U} J^2(P, \mathbf{R})_p \subset J^2(P, \mathbf{R})$$

Define a function from U into $\mathbf{R}^r$ by

$$\theta_p(q) = \theta(q) - \theta(p)$$

The function θ_p defines co-ordinates on U with the property that $\theta_p(p) = 0$. Then a basis of $J^2(P, \mathbf{R})_q$ is given by the 2-jets of the functions

$$1, \theta^i_q, \theta^j_q \theta^k_q$$

for $1 \leq i \leq r$ and $1 \leq j < k \leq r$. Here 1 denotes the constant function on u whose value is 1.

If f is a function defined about p then its 2-jet is an element of the vector space $J^2(P, \mathbf{R})_p$ so it can be expanded in the basis we have given. This expansion is

$$j^2_p(f) = f(p).1 + \sum_i \frac{\partial f}{\partial \theta^i}(p)\theta_p + \frac{1}{2}\sum_{j,k} \frac{\partial^2 f}{\partial \theta^j \partial \theta^k}(p)\theta^j_p \theta^k_p$$

Notice that it makes no difference whether we calculate our partial derivatives with respect to the θ co-ordinates or the θ_p co-ordinates.

As q varies this basis defines a local frame for $J^2(P, \mathbf{R})$ and we leave it as an exercise for the reader to show that these local frames make $J^2(P, \mathbf{R})$ a vector bundle in a manner analogous to the tangent bundle.

Before considering Taylor series we need to know a little more of the structure of these spaces of jets. It is an easy exercise to show that $J^\infty_p(P, \mathbf{R})$ shares most of the properties of $C^\infty(P, \mathbf{R})$ the space of all smooth functions on P. In particular it is a vector space; one can add jets and multiply them by scalars by doing the same things to representative functions in the jet equivalence class. Notice that it has to be checked that the result is independent of the choice of function. Similarly it is possible to show that $J^\infty_p(P, \mathbf{R})$ is an algebra; one can multiply jets by multiplying representative functions for them.

If f and g are in the same r-jet and f vanishes at p so also must g. So it makes sense to talk of the jets in $J^\infty_p(P, \mathbf{R})$ which vanish at p. These form an ideal in $J^\infty_p(P, \mathbf{R})$ and we denote it by I. From this ideal we can form the ideals I^p for any p. This is the ideal generated by all pth products of elements of I. Equivalently it is the collection of all jets vanishing to order $k-1$ at k. Clearly these ideals are all nested:

$$I = I^1 \subset I^2 \subset \ldots \subset I^p \subset \ldots J^\infty_p(P, \mathbf{R})$$

We want to understand the quotients I^k/I^{k-1}. Consider the same thing when $P = \mathbf{R}^n$ and $p = 0$. Then functions are jet equivalent when they have the same Taylor expansion so Taylor expansion defines a map

$$T: J^\infty(\mathbf{R}^n,\ \mathbf{R}) \to P(\mathbf{R}^n)$$

Here $P^k(\mathbf{R}^n)$ is the vector space of all homogeneous polynomials of degree k on $\mathbf{R}^n$ and $P(\mathbf{R}^n) = \oplus_{k\geq 0} P^k(\mathbf{R}^n)$ the algebra of all formal series. A theorem of Borel states that this map is onto Golubitsky and Guillemin (1973). That is for every formal series there is a smooth function whose Taylor series is that formal series. Notice that we are making no claims here about convergence.

In this picture the ideal I is

$$I = \oplus_{j\geq 1} P^j(\mathbf{R}^n)$$

and the ideal I^k is

$$I^k = \oplus_{j\geq k} P^j(\mathbf{R}^n)$$

Hence we have

$$I^k/I^{k-1} = P^k(\mathbf{R}^n) \tag{9.2.1}$$

If we choose co-ordinates on P about the point p then we expect a similar thing to be true but how do we interpret $P^k(\mathbf{R}^n)$? We can replace $\mathbf{R}^n$ by T_pP so it remains to be able to define polynomials on a vector space. In the case of $\mathbf{R}^2$ for example monomials of degree k look like $(x^1)^k$, $(x^1)^{k-1}\,x^2, \ldots$ etc. where $x^1,\ x^2: \mathbf{R}^2 \to \mathbf{R}$ are the co-ordinate functions. A monomial of degree 1 therefore is a linear function and a monomial of degree k is a k-fold product of linear functions. On a general vector space V we can define $S^k(V^*)$ to be the vector space of all symmetric multi-linear maps

$$\underbrace{V \times \ldots \times V}_{k \text{ times}} \to \mathbf{R}$$

A map is symmetric and multilinear if it is linear in each factor and symmetric under permutation of the variables. The generalisation of (9.2.1) is therefore

$$I^k/I^{k-1} = S^k(T_p^*P)$$

Another way of viewing this result is to notice that $J^\infty_p(P,\mathbf{R})/I^k$ can be identified with $J^{k-1}_p(P,\mathbf{R})$. Moreover there is a natural map from $J^k_p(P,\mathbf{R})$ to $J^{k-1}_p(P,\mathbf{R})$ that just sends the k-jet of a function to its $k-1$-jet. It follows from a little linear algebra that the subpace of elements mapped to zero by this map is I^k.

A little terminology may help clarify the situation. If W is a vector space we say it is *graded* if it has subspaces W_i and

$$W = \bigoplus W_i$$

We say W is *filtered* if it has subspaces V_i with

$$\ldots \subset V_i \subset V_{i+1} \subset \ldots \subset W$$

and

$$W = \cup_i V_i$$

If we have a grading we can define a filtering by letting

$$V_i = \oplus_{j\geq i} W_j$$

Also if we have a filtering we can form a graded vector space

$$\oplus(V_i/V_{i-1})$$

but this is not the vector space W.

So in $\mathbf{R}^n$ we have that $J^\infty_p(\mathbf{R}^n)$ is a graded space; graded by the monomials of various degrees. However when we pass to a general manifold $J^\infty_p(P,\mathbf{R})$ is only a filtered space, filtered by the I_p. The corresponding graded space is

$$\oplus_{k=0}^\infty S^k(T^*_p(P))$$

Example 9.2.2 *2-frames.* Let us denote by $J^2(P,\mathbf{R}^n)_p$ the set of all 2-jets at p of functions from P to $\mathbf{R}^n$. Inside here we can consider those which are actually 2-jets of co-ordinate systems in some open set about p and which map p to 0 in $\mathbf{R}^n$. The inverse function theorem tells us that $\theta\colon P \to \mathbf{R}^n$ defines a co-ordinate system in some neighbourhood of p precisely when the tangent at p, $T_p(\theta)$ is invertible. Note that this condition depends only on the 1-jet equivalence class of θ at p, i.e. it depends only on the first derivatives of θ at p. The jets of this type are what we define to be 2-frames in section 9.3.

9.3 Taylor series and co-ordinate strings

9.3.1 *Taylor series*

In the calculations of Amari it is clear that Taylor series play an important role. We want to consider the definition of Taylor series on a manifold.

A Taylor series for a function $f: \mathbf{R}^n \to \mathbf{R}$ about zero is a formal series of polynomials with coefficients given by the higher derivatives of f. It is important to reiterate that we *do not* require that the series converges in any sense; we are interested only in the higher derivatives. In $\mathbf{R}^n$ we can think of the operation of forming the Taylor series as a function

$$T: J_0^\infty(\mathbf{R}^n) \to P(\mathbf{R}^n)$$

which we will call a Taylor map. Notice that the Taylor map is an algebra *isomorphism* and preserves the filterings of both spaces.

If we replace $\mathbf{R}^n$ by an n-dimensional vector space V we wouldn't expect much of a change in the above discussion except that, as we saw in section 8.1 we should replace polynomials by symmetric functions. The Taylor map is then an algebra isomorphism

$$J_0^\infty(V) \to S(V^*)$$

where

$$S(V^*) = \sum_k S^k(V^*)$$

Notice that we are now regarding the higher derivatives of a function as elements of $S(V^*)$.

For a function on a manifold, just as for rates of change of vector fields, there are lots of different ways of defining Tayor series so we shall define the Taylor map by listing the properties we expect it to have. A Taylor map (at p) is defined to be an isomorphism

$$T: J_p^\infty(P) \to \bigoplus_{k\geq 0} S^k(T_p^* P) = S(T^* P)$$

such that

$$T^0(f) = f(p)$$

$$T^1(f) = df$$

and

$$T(fg) = T(f)T(g)$$

By $T^k(f)$ we mean the component of $T(f)$ in $S^k(T_p^*P)$. The first and second of these conditions ensure that the zeroth and first derivative, which we can define already, are what we expect them to be and the third condition, of course, just says that T is an algebra homomorphism. Notice that the first condition means that T will map I to $\oplus_{i\geq 1}S^i(T_p^*P)$. Because T is an algebra isomorphism it now follows that T will map I^k to $\oplus_{i\geq k}S^i(T_p^*P)$, that is T preserves the filtration on both side.

Let us consider some examples.

Example 9.3.1 If ∇ is a connection on P then repeated application of the connection defines a k linear map $\nabla^{(k-1)}df(p)$ on the tangent space at p and if we let $T^k(f)$ be the symmetrisation of this map for $k > 1$ and $T^0(f) = f(p)$ and $T^1(f) = df(p)$ we obtain a Taylor map.

Example 9.3.2 If $\varphi = (\varphi^1, \ldots, \varphi^n)$ are co-ordinates then Taylor expanding *relative* to these defines a Taylor map T_φ with

$$(T_\varphi^k f)(p) = \sum_{|\alpha|=k} D_\varphi^\alpha f(p)(d\varphi^1)^{\alpha_1} \ldots (d\varphi^n)^{\alpha_n}$$

Here $\alpha = (\alpha_1, \ldots, \alpha_n)$ is a multi-index,

$$|\alpha| = \alpha_1 + \ldots + \alpha_n \quad \text{and} \quad D_\varphi^\alpha f = \frac{\partial^{|\alpha|} f}{(\partial\varphi^1)^{\alpha_1} \ldots (\partial\varphi^n)^{\alpha_n}}$$

Example 9.3.3 *The exponential map.* Recall the definition of the exponential map of a connection and the associated normal co-ordinates. The Taylor map defined by these co-ordinates and that defined by the connection ∇ are the same.

So far we have considered the notion of a Taylor map at a point of a manifold. We should also like to define the idea of a choice of Taylor map that varies smoothly over the manifold. The definition is easy enough. If we are given a Taylor map at every point of P we say it is a smooth Taylor map on P if for every smooth function f on P the symmetric tensors $T^k(f)$ are smooth. Usually we just say Taylor map rather than smooth Taylor map.

There is another way of defining a Taylor series that relates to the strings of Barndorff-Nielsen, Murray (1986).

9.3.2 *Co-ordinate strings*

We have seen in example 9.3.2 that every choice of co-ordinates on P gives rise to Taylor map. It is not hard to see that if θ and φ are co-ordinates then $T_\theta = T_\varphi$ if and only if there is an invertible matrix $X = X^i_j$ such that θ and $X\varphi$ are infinite jet equivalent at p, where

$$(X\varphi)^i = \sum_j X^i_j \varphi^j$$

We call such an equivalence class an (infinite) co-ordinate string at p.

What interests us is that every Taylor map arises in this way. To see this choose a basis ω^i for $T^*_p P$ and functions θ^i such that

$$T^1(\theta^i) = d_p\theta^i = \omega^i$$

It is clear that as a map $\theta: P \to \mathbf{R}^n$ this maps p to zero and has invertible derivative. Hence, by the implicit function theorem, it is locally a set of co-ordinates. It remains to show that $T = T_\theta$. But from the definition $T_\theta(1) = 1 = T(1)$ and $T(\theta^i) = T_\theta(\theta^i)$ and these $r+1$ elements generate $J^\infty_p(P)$. Because T and T_θ are both algebra isomorphisms and the algebra is generated by the functions 1 and θ^i it follows that T and T_θ are equal.

So having a Taylor map is equivalent to having a co-ordinate string.

It follows that if f is a function and φ is a co-ordinate string then we can construct the Taylor expansion of f relative to the Taylor map induced by the φ. This defines a sequence of symmetric tensors $T^k_\varphi(f)$ which are precisely those obtained by *intertwining*, in the sense of Barndorff-Nielsen and Blæsild (1987), the connection string φ with the scalar string f.

We consider now one final definition of a co-ordinate string that is useful in examples. Let $\theta: P \to T_pP$ be a map with $\theta(p) = 0$ and such that

$$T_p(\theta): T_pP \to T_0(T_pP) = T_pP$$

is the identity. Then if we choose a basis of T_pP the component functions of θ define co-ordinates about p and hence a co-ordinate string at P. It is easy to check that the infinite jet class of θ precisely determines a co-ordinate string. Given the infinite jet class of such a θ note that it defines an algebra isomorphism

$$j^\infty(\theta): J^\infty_0(T_pP) \to J^\infty_p(P)$$

by composing functions on T_pP with θ. On the other hand, as a vector space T_pP has a Taylor map

$$J_0^\infty(T_pP) \rightarrow S(T_p^*P)$$

defined by its linear co-ordinates. Composing this with the inverse of $j^\infty(\theta)$ defines the Taylor map defined by θ.

Example 9.3.4 *The natural co-ordinate string of a family.* Consider our statistical example. If we choose a point p in P then we can identify R_Ω with $\mathcal{M}$ and this also identifies T_pP with a subspace of R_Ω. Orthogonal projection from R_Ω to T_pP now defines a map

$$P \subset \mathcal{M} \simeq R_\Omega \rightarrow T_pP$$

It is easy to check that this map sends p to zero and is the identity on the tangent spaces. Hence it defines a co-ordinate string called the natural co-ordinate string of the family. Notice that instead of this projection we could identify $\mathcal{P}$ with $R_\Omega/\mathbf{R}.1$ then identify $R_\Omega/\mathbf{R}.1$ with the set of random variables whose expectation with respect to p is zero and finally project to T_pP. The result would be the same. In the latter case it is clear that if P is an exponential family then inside $R_\Omega/\mathbf{R}.1$ P and T_pP coincide so that this map is that which defines that canonical co-ordinates on P. Hence the choice of name. (See also 6.7.2.)

To calculate this map explicitly let us choose local co-ordinates θ and note that the orthogonal projection of a random variable f is

$$\pi_p(f) = \sum_{i,j} g^{ij}(p)E_p(fl_i)\frac{\partial}{\partial\theta^j}$$

It follows that the natural co-ordinate string is

$$q \mapsto g^{ij}(p)E_p((\log\ q - \log\ p)\frac{\partial\ell}{\partial\theta^i})$$

The natural co-ordinate string has been used by McCullagh and Cox (1986) in their calculations of the Bartlett adjustment. Their Möbius derivatives are just the Taylor coefficients of the log likelihood function in the natural co-ordinate string.

9.3.3 *Components of a co-ordinate string*

In case the reader should be brave enough to want to work with co-ordinate strings in local co-ordinates we give here the definition of the components of a co-ordinate string.

If $\theta = (\theta^1, \ldots, \theta^n)$ is a co-ordinate system in a neighbourhood of p and φ is a co-ordinate string at p then we can define the collection of numbers

$$M^i_{i_1 \ldots i_k}(p) = \frac{\partial \theta^i}{\partial \varphi^j}(p) \frac{\partial^k \varphi^i}{\partial \theta^{i_1} \ldots \partial \theta^{i_k}}(p)$$

We call these numbers the components of φ with respect to θ. Clearly the determine the co-ordinate string φ. Conversely given a collection of numbers like this which are symmetric in their lower indices and such that M^i_j is invertible then we can define a co-ordinate string for which these are the components. This follows from the lemma of Borel and the implicit function theorem. Such a collection of numbers is called a structurally symmetric connection string in the general theory of strings developed by Barndorff-Nielsen and Blæsild (1987).

9.3.4 *Flatness of co-ordinate strings*

If $(\varphi^1, \ldots, \varphi^n)$ is a co-ordinate system defined on a set U it determines a co-ordinate string over the set U by

$$\varphi_q = (\varphi^1_q, \ldots, \varphi^n_q) = (\varphi^1 - \varphi^1(q), \ldots, \varphi^n - \varphi^n(q))$$

Some but not all families of co-ordinate strings arise in this way. Such co-ordinate strings are called flat.

We will give a criterion here for flatness of a co-ordinate string. The reader uninterested in this topic is advised to skip this section. The reader familiar with principal bundles is advised to look at the treatment of Koboyashi (1961) of which this is a co-ordinate version. We define a collection of 1-forms w^r for $r \geq 2$ whose vanishing is equivalent to the flatness of the co-ordinate string. These 1-forms are invariants of the string and are related to the canonical 1-forms defined by Koboyashi. We refer the reader to Murray (1986) for the proof that vanishing of these 1-forms is equivalent to flatness.

Let φ_q be a co-ordinate string defined in a neighbourhood of a point p. Then we can Taylor expand φ_q as

$$\varphi_q^i = (\varphi_p^i - \varphi_p^i(q)) + \frac{1}{2!} M^i_{jk}(q)(\varphi_p^j - \varphi_p^j(q))(\varphi_p^k - \varphi_p^k(q)) + \ldots$$

where we use the freedom in the definition of the string to choose φ_q so that the coefficient of the first term $M^i_j = \delta^i_j$. Flatness is just the property that $M^i_{j_1 \ldots j_k}(q) = 0$ for all $j_1, \ldots, j_k$ and for all q in the neighbourhood of p in question. The 1-form w^r is defined by

$$w^r(p) = d(M^j_{i_1 \ldots i_r})(p)\; d\varphi_p^i(p) \otimes \ldots \otimes d\varphi_p^{i_r}(p) \otimes \frac{\partial}{\partial \varphi_p^j}\,(p)$$

and is a 1-form with values in $S^r(T_p^* P) \otimes T_p P$, the rth symmetric tensor power of the cotangent space tensored with the tangent space, because the indices $i_1, \ldots, i_r$ are symmetric.

Example 9.3.5 In the notation of McCullagh and Cox (1986) for the natural co-ordinate string we have

$$\varphi_q^i = (\varphi_p^i - \varphi_p^i(q)) + \frac{1}{2!}\,\nu^{ij}(q)\nu_{j,kl}(q)(\varphi_p^k - \varphi_p^k(q))(\varphi_p^l - \varphi_p^l(q)) + \ldots$$

The 1-forms w^r at p are therefore

$$w^r(p) = \sum \frac{\partial}{\partial \varphi_p^i}\left[\nu^{jk}\; \nu_{k,l_1 \ldots l_r}\right]|_p \;\; \frac{\partial}{\partial \varphi_p^j} \otimes d\varphi_p^i \otimes d\varphi_p^{l_1} \ldots \otimes d\varphi_p^{l_r}$$

Their vanishing amounts to the vanishing of the

$$\begin{aligned}
&\frac{\partial}{\partial \varphi_p^i} \nu^{j,k}(p)\; \nu_{k,l_1 \ldots l_r}(p) \\
&\quad = -\nu^{j,l}\; \nu^{r,k}\{v_{i,l,r} + \nu_{il,r} + \nu_{l,ir}\}\nu_{k,l_1 \ldots l_r} \\
&\qquad + \nu^{j,k}\; \nu_{i,k,l_1 \ldots l_r} + \nu^{j,k}\; \nu_{ik,l_1 \ldots l_r} + \nu^{j,k}\; \nu_{i,kl_1 \ldots l_r} \\
&\quad = \nu^{j,k}(p)\; \nu_{i,k,l_1 \ldots l_r}(p) + \nu^{j,k}(p)\; \nu_{ik,l_1 \ldots l_r}(p)
\end{aligned}$$

Here the ν's are the cumulants of the derivatives of the log likelihood functions, for example

$$\nu_{i,j,kl}(q) = E_q\left(\frac{\partial l}{\partial \varphi_p^j}(q) \frac{\partial l}{\partial \varphi_p^j}(q) \frac{\partial^2 l}{\partial \varphi_p^k \partial \varphi_p^l}\,(q)\right)$$

and we have used the fact that $\nu_{i,l_1 \ldots l_s}(p) = 0$ for $s \geq 2$.

9.3.5 *Co-ordinate strings and the infinite frame bundle*

Recall the definition of the infinite frame bundle in Chapter 8. The fibre $\mathcal{F}_p$ at p is the set of all infinite jets of co-ordinate functions at p. We noted that this was acted on by $\mathcal{G}^\infty(n)$. Now amongst the diffeomorphisms of $\mathbf{R}^n$ there are the invertible linear transformations. So the group $GL(n)$ is a subgroup of $\mathcal{G}^\infty(n)$ and hence also acts on $\mathcal{F}_p$. We see from the definition of a co-ordinate string that the quotient by this $GL(n)$ action defines the set of all co-ordinate strings at p. We can form the union of all these sets to obtain the bundle of all co-ordinate strings which is the quotient of $\mathcal{F}^\infty(P)$ by $GL(n)$. A co-ordinate string on P is just a section of this bundle.

9.4 Strings and phylon representations

We want to say something briefly about the general theory of strings which is a topic for a book in its own right. We shall approach these from the point of view of principal bundles. For further references to the extensive literature on strings and a survey relating what we do here to the co-ordinate approach known as the theory of **D**-matrices we refer the reader to Barndorff-Nielsen et. al (1992) from which some of this material is taken.

A string is a generalisation of a tensor in the sense that in the transformation formula (8.7.1) we should allow higher derivatives of the change of co-ordinate function. Consider for instance the second jet bundle $J^2(P)$. If we choose local co-ordinates θ then a basis for $J^2(P, \mathbf{R})_p$ is given by 2 jets of the functions

$$1, \quad \theta^i, \quad \text{and} \quad \theta^j\theta^k$$

where the i, j and k run from 1 to r with $j \leq k$. Then a section f of $J^2(P)$ can be expanded in this basis as

$$f = f(\theta).1 + f(\theta)_i\theta^i + f(\theta)_{jk}\theta^j\theta^k$$

In co-ordinates f is determined by the string of numbers

$$(f(\theta), f(\theta)_i, f(\theta)_{jk})$$

If we change co-ordinates the brave reader can calculate that that the transformation rule is summarised by the following formulae:

$$\begin{pmatrix} f(\theta) \\ f(\theta)_i \\ f(\theta)_{jk} \end{pmatrix} = \begin{pmatrix} 1 & 0 & 0 \\ 0 & \frac{\partial \chi^a}{\partial \theta^i} & 0 \\ 0 & \frac{\partial^2 \chi^a}{\partial \theta^j \partial \theta^k} & \frac{\partial \chi^b}{\partial \theta^j}\frac{\partial \chi^c}{\partial \theta^k} \end{pmatrix} \begin{pmatrix} f(\chi) \\ f(\chi)_a \\ f(\chi)_{bc} \end{pmatrix}$$

From the point of view of principal bundles a string is a section of a vector bundle associated to the infinite frame bundle by a representation of $\mathcal{G}^\infty(n)$. The study of the algebraic properties of strings reduces therefore to the study of the representations of the group $\mathcal{G}^\infty(n)$. To be able to develop a reasonable theory we restrict attention to a particular set of representations called *phylon representations* by Barndorff-Nielsen et al (1992) and also studied in Terng (1978). This set of representations has the property that it is closed under various natural operations. For instance if V and W are phylon representations then so also is $V \oplus W$ and $V \otimes W$ and if W is a subrepresentation of a phylon representation V then W is a phylon representation and so also is the quotient space V/W. In the remainder of this chapter we will explain what phylon representations are and discuss some of their structure. In simple cases we shall be able to classify all phylon representations. Terng (1978) gives a far more general classification theorem in terms of Lie algebra cohomology which we shall not attempt to explain here.

9.4.1 *Phylon representations*

We will restrict attention to the finite-dimensional phylon groups $\mathcal{G}_T$ which are the groups of diffeomorphisms of the origin identified up to T-jet equivalence and to finite-dimensional representation spaces V. This is not as big a restriction as it may seem. Terng (1978) proves that any finite-dimensional representation that is continuous, in a reasonable sense, is really the composition of a representation from some $\mathcal{G}_T$ with the projection $\mathcal{G}_\infty \to \mathcal{G}_T$. There are infinite-dimensional representations of $\mathcal{G}_\infty$ that do not arise in this way, for instance the action on functions where if ϕ is the germ of a diffeomorphism and f a function we have $\phi.f = f \circ \phi^{-1}$, but we shall not consider these here. We would expect that the theory of infinite-dimensional representations of $\mathcal{G}_\infty$ would be complicated. It is worth noting that Terng in fact proves the stronger result that any continuous finite-dimensional representation of the group of germs of diffeomorphisms of $\mathbf{R}^n$ fixing the origin factors through some $\mathcal{G}_T$.

If V is a representation of $\mathcal{G}_T$ we shall call it *algebraic* if in any basis of V the matrices representing the elements such as f in (8.5.1) have entries that are polynomial functions of the $f^i_{j_1 j_2 \ldots j_k}$ and $\det(f^i_{j_1})^{-1}$. This is the definition used in the theory of algebraic groups; see for instance Humphreys (1981). We shall call a finite-dimensional complex representation of $\mathcal{G}_T$ a phylon representation if it is an algebraic representation. The general theory of algebraic groups and their representations guarantees that phylon representations are well behaved in the sense we discussed above. This is also in Terng (1978).

Before turning to the theory of phylon representations we need to consider some of the structure of the infinite phylon group $\mathcal{G}_\infty(n)$. For simplicity we shall now drop the reference to n in the notation. The map that associates to an infinite jet its T-jet defines a group homomorphism

$$\mathcal{G}_\infty \to \mathcal{G}_T$$

whose kernel is a subgroup denoted by denoted by $\mathcal{G}^{(T)}$. If $T > S$ then we also have a homomorphism

$$\mathcal{G}_T \to \mathcal{G}_S$$

onto $\mathcal{G}_S$ and we shall denote its kernel by $\mathcal{G}_T^{(S)}$. Of particular interest in what follows is the map

$$\mathcal{G}_T \to \mathcal{G}_1 = GL(n)$$

whose kernel is the subgroup $\mathcal{G}_T^{(1)}(d)$ of power series of the form

$$\delta^i_j x^j + \frac{1}{2!} f^i_{j_1 j_2} x^{j_1} x^{j_2} + \ldots + \frac{1}{T!} f^i_{j_1 j_2 \ldots j_T} x^{j_1} x^{j_2} \ldots x^{j_T}$$

As well as the projection from $\mathcal{G}_T$ to $GL(d)$, there is an inclusion of $GL(d)$ in $\mathcal{G}_T$ which sends $X \in GL(d)$ with components X^i_j to the power series

$$X^a_i x^i$$

Thus we can regard $GL(d)$ as a subgroup of $\mathcal{G}_T$. It is easy to see that composition in $\mathcal{G}_T$ gives a bijection

$$\mathcal{G}_T^{(1)} \times GL(d) \to \mathcal{G}_T$$

Hence it defines a group product on

$$\mathcal{G}_T^{(1)} \times GL(d)$$

This is not the direct product but a semi-direct product defined as follows. If we start with a pair (f, X) and (g, Y) of elements of the direct product then they map to the elements fX and gY in $\mathcal{G}_T$. The product of these in $\mathcal{G}_T$ is $fXgY = f(XgX^{-1})XY$. Notice that XgX^{-1} is an element of $\mathcal{G}_T^{(1)}(d)$ so that this product corresponds to $(fXgX^{-1}, XY)$ as an element of $\mathcal{G}_T^{(1)} \times GL(d)$. We shall be interested in the structure of phylon representations relative to the semi-direct product decomposition of $\mathcal{G}_T$.

Example 9.4.1 The fundamental example of a phylon representation is the representation of $\mathcal{G}_T$ on $\mathcal{C}_T$ the space of power series of degree less than or equal to T with no constant term. This is a direct sum of the various homogeneous polynomials so we have

$$\mathcal{C}_T = \bigoplus_{k=1}^{T} S^k(\mathbf{R}^n)^*$$

where $S^k(\mathbf{R}^n)^*$ denotes the symmetric product of k copies of the dual space of $\mathbf{R}^n$ or the space of polynomials homogeneous of degree k. This decomposition realises $\mathcal{C}_T$ as a direct sum of irreducible representations of $GL(n) \subset \mathcal{G}_T$. With respect to this decomposition any element of $\mathcal{G}_T$ can be represented as a block matrix. The entry in the ith row and jth column is an operator from $S^j(\mathbf{R}^n)^*$ to $S^i(\mathbf{R}^n)^*$. The elements of $GL(n)$ are of course represented by block diagonal matrices. Every element is represented by an upper triangular matrix and that $\mathcal{G}_T^{(1)}$ is represented by block matrices having identity matrices along the diagonal.

The most important fact about this situation is Theorem 3.8 of Terng (1978). This says that every phylon representation is of this form. So we start with a finite-dimensional algebraic representation V of $\mathcal{G}_T$ and decompose it as

$$V = \bigoplus_{i=1}^{r} V_i$$

where each V_i is an irreducible representation of $GL(n) \subset \mathcal{G}_T$. This much follows from the representation theory for algebraic groups. Terng's theorem now ways that we can order the V_i in such a way that the elements of $\mathcal{G}_T^{(1)}$ are represented by block upper-triangular matrices with identity matrices along the diagonal.

Consider what this tells us about the structure of phylon representations. Let $\chi : \mathcal{G}_T \to GL(V)$ be an indecomposable phylon representation and let $V = V_1 \oplus \cdots \oplus V_k$ be the decomposition into $GL(n)$ irreducibles that we have constructed above. We shall call the ordered set $(V_1, \ldots, V_k)$ of $GL(n)$ modules the *type* of the representation and the number k here the *rank* of the representation. Then if fX is an element of the phylon group with f in $\mathcal{G}_T^{(1)}$ and X in $GL(n)$ then the operator $\chi(X)$ in $GL(V)$ can be written as

$$\chi(X) = \begin{pmatrix} \chi_1(X) & 0 & \ldots & 0 \\ 0 & \chi_2(X) & \ldots & 0 \\ \vdots & \vdots & \ddots & \vdots \\ 0 & 0 & \ldots & \chi_k(X) \end{pmatrix}$$

and the operator $\chi(f)$ as

$$\chi(f) = \begin{pmatrix} 1 & \chi_{12}(f) & \ldots & \chi_{1k}(f) \\ 0 & 1 & \ldots & \chi_{2k}(f) \\ \vdots & \vdots & \ddots & \vdots \\ 0 & 0 & 0 & 1 \end{pmatrix}$$

so that the product $\chi(fX) = \chi(f)\chi(X)$ is

$$\chi(fX) = \begin{pmatrix} \chi_1(X) & \chi_{12}(f)\chi_2(X) & \ldots & \chi_{1k}(f)\chi_k(X) \\ 0 & \chi_2(X) & \ldots & \chi_{2k}(f)\chi_k(X) \\ \vdots & \vdots & \ddots & \vdots \\ 0 & 0 & \ldots & \chi_k(X) \end{pmatrix}$$

where $\chi_i(X)$ is an operator on V_i, that is, $\chi_i(X) : V_i \to V_i$ is a linear map and $\chi_{ij}(f) : V_j \to V_i$ is also a linear map. These linear maps satisfy $\chi_{ii}(1) = 1$ and $\chi_{ij}(1) = 0$ for $i \neq j$.

Because χ is a representation of the phylon group we must have

$$\chi(fXgY) = \chi(fX)\chi(gY)$$

and using the matrix formulation above we deduce firstly that

$$\chi_i(X)\chi_i(Y) = \chi_i(XY)$$

so that the $\chi_i : GL(n) \to GL(V_i)$ are representations, and secondly that

$$\chi_{ij}(XgX^{-1}) = \chi_i(X)\chi_{ij}(g)\chi_j(X)^{-1} \tag{9.4.1}$$

and

$$\chi_{ij}(fg) = \chi_{ij}(g) + \chi_{i,i+1}(f)\chi_{i+1,j}(g) + \chi_{i,i+2}(f)\chi_{i+2,j}(g) + \\ \dots + \chi_{i,j-1}(f)\chi_{j-1,j}(g) + \chi_{ij}(f) \tag{9.4.2}$$

where we have inserted some commas to clarify the notation. Moreover the same calculation shows that if we start with the χ_i and χ_{ij} satisfying these three conditions then the χ defined by (9.4.2) is a representation. Notice that χ_{ij} is a map

$$\chi_{ij} : \mathcal{G}_T^{(1)} \to \text{Hom}(V_j, V_i) \simeq V_j^* \otimes V_i$$

and that $GL(n)$ acts on both sides of this map. The condition (9.4.1) just says that χ_{ij} commutes with the action of $GL(n)$. If all the χ_{ij} are zero, so that $\mathcal{G}_T^{(1)}$ acts trivially we shall call this representation the *trivial* phylon representation of type $(V_1, \dots, V_k)$.

Example 9.4.2 Let us consider a simple case, when $T = 2$. The elements of this phylon group are of the form

$$\alpha = \alpha^i_j x^j + \alpha^i_{jk} x^j x^k$$

and the functions in $\mathcal{C}_2$ are of the form

$$\phi = \phi_i x^i + \phi_{ij} x^i x^j$$

The action of the element α^{-1} is given by

$$\begin{aligned}(\alpha^{-1}\phi)(x) =& \phi(\alpha(x)) \\ =& \phi(\alpha^i_j x^j + \alpha^i_{jk} x^j x^k) \\ =& \phi_i(\alpha^i_j x^j + \alpha^i_{jk} x^j x^k) + \phi_{il}(\alpha^i_j \alpha^l_k x^j x^k) \\ =& (\phi_i \alpha^i_j) x^j + (\phi_i \alpha^i_{jk} + \phi_{il} \alpha^i_j \alpha^l_k) x^j x^k\end{aligned}$$

where we use the fact that terms of order higher than 2 are discarded.

From this we discover that relative to the splitting

$$V_1 = \{\phi_{ij} x^i x^j\} = S^2 \mathbf{R}^{n*}$$

and

$$V_2 = \{\phi_i x^i\} = \mathbf{R}^{n*}$$

we have

$$\chi(\alpha) = \begin{pmatrix} \alpha^i_l \alpha^j_k & \alpha^i_{jk} \\ 0 & \alpha^i_l \end{pmatrix}$$

Consider in general the classification of representations of $\mathcal{G}_2$ which are of rank 2. They are therefore of the form

$$V = V_1 \oplus V_2.$$

From the theory we have developed above we see that to define a representation of $\mathcal{G}_2$ we have to choose the two representations $\chi_i : GL(n) \to GL(V_i)$ and a map

$$\chi_{12} : \mathcal{G}_2^{(1)} \to V_2^* \otimes V_1$$

such that

$$\chi_{12}(XgX^{-1}) = \chi_1(X)\chi_{12}(g)\chi_2(X^{-1})$$

and

$$\chi_{12}(fg) = \chi_{12}(f) + \chi_{12}(g)$$

Notice that $\mathcal{G}_2^{(1)}$ is an abelian group. Indeed the obvious bijection $\mathcal{G}_2^{(1)} \simeq \mathbf{R}^n \otimes S^2\mathbf{R}^{n*}$ is an isomorphism of groups if we put the additive structure on the second space. All this is saying is that if $\alpha = \delta^i_j x^j + \alpha^i_{jk} x^j x^k$ and $\beta = \delta^i_j x^j + \beta^i_{jk} x^j x^k$ then $\alpha\beta = \delta^i_j x^j + (\alpha^i_{jk} + \beta^i_{jk}) x^j x^k$. Hence

$$\chi_{12} : \mathbf{R}^n \otimes S^2\mathbf{R}^{n*} \to V_2^* \otimes V_1$$

is an additive, continuous map commuting with the group action on either side. Recall that it is straightforward to show that an additive continous map between vector spaces is linear. So the classification of the phylon representations of $\mathcal{G}_2$ of rank 2 reduces to the study of these linear maps χ_{12}. In the example of $\mathcal{C}_2$ this is essentially the identity map.

Of course, we are really interested in isomorphism classes of phylon representations. In general, if two representations χ and χ' of the phylon group are isomorphic this means that we have an isomorphism of vector spaces

$$\Phi : V \to V$$

such that

$$\chi(fX)\Phi(v) = \Phi(\chi'(fX)(v))$$

for all v in V. In the case that $V = V_1 \oplus V_2$ then we can put Φ into 2×2 block form. As Φ, in particular, intertwines the $GL(n)$ actions we have, by Schur's lemma, Humphreys (1981), that

$$\Phi = \begin{pmatrix} \alpha & 0 \\ 0 & \beta \end{pmatrix}$$

where α and β are *multiples of the identity matrix.*

In particular if we choose arbitrary α and β this intertwines representations with $\chi_{12} = \chi$ and $\chi_{12} = (\alpha/\beta)\chi$. So multiplying χ_{12} by a non-zero scalar gives rise to an isomorphic representation.

We have now shown that the isomorphism classes of phylon representations of $\mathcal{G}_2$ of type (V_1, V_2) are in 1-1 correspondence with $GL(n)$ equivariant maps

$$\chi_{12} : \mathbf{R}^n \otimes S^2\mathbf{R}^{n*} \to V_2^* \otimes V_1$$

taken up to scale. That is, if $X(V_1, V_2)$ denotes the set of all such $GL(n)$ equivariant maps and we let $\mathbf{C}^\times$ act by scalar multiplication, the isomorphism classes of phylon representations are in $1-1$ correspondence with the set of $\mathbf{C}^\times$ orbits denoted by $X(V_1, V_2)/\mathbf{C}^\times$. We can, in principle, use this information to classify all the phylon representations of this type. There are combinatorial formulae (such as Young tableaux) which show how to decompose a tensor product of irreducible $GL(n)$ representations into irreducibles. In particular we have that $\mathbf{R}^n \otimes S^2\mathbf{R}^{n*}$ is a sum of two non-isomorphic irreducibles W_1 and W_2 say and $V_2^* \otimes V_1$ is a sum of r irreducibles U_i say. Then Schur's lemma tells us that the only possible $GL(n)$ equivariant maps χ_{12} are those that map each W_i isomorphically to a U_j. This means that if there are p_i of the U_j isomorphic to W_i then

$$X(V_1, V_2) \simeq \mathbf{C}^{p_1+p_2}$$

where $\mathbf{C}^\times$ acts on this latter space by scalar multiplication of vectors. The various possibilities can now be computed by investigating the orbit space. The orbit of a vector is either $\{0\}$ if the vector is 0 or a line without zero if the vector is non-zero. It follows that the space of orbits is a point corresponding to zero and $\mathbf{CP}_{p_1+p_2-1}$ the space of all lines in $\mathbf{C}^{p_1+p_2}$. The trivial representation corresponds to the point. In particular, for $V_1 = \mathbf{R}^{n*}$ and $V_2 = S^2\mathbf{R}^{n*}$ we find that $X(V_1, V_2) \simeq \mathbf{C}^2$ and we obtain the 2-sphere $\mathbf{C}P_1$.

This example and the more general case of $\mathcal{G}_T$ acting on a V which has two $GL(n)$ irreducible factors are discussed in Terng (1978), the latter case following from Terng's Theorem 3.14 which gives a general classification scheme in terms of group orbits on Lie algebra cohomology spaces. While they provide an answer to the question of classifying phylon representations these cohomology spaces and the orbit structure are difficult to calculate in general, so from a practical viewpoint the specific calculations above are probably the best one can do. The interested reader may like to consult Terng (1978) for further details.

9.5 Remarks for Chapter 9

Remark 9.1 Our view of strings here has been that they are generalisations of tensors. This is in keeping with their historical origins and development in the work of Foster (1958), (1989) and references therein, where they were called new-tensors.

Remark 9.2 While we have presented here an overview of the structure of the theory of strings we have not presented the many elegant geometric constructions of particular examples, in particular the solution by Jupp (1992) of the problem of finding a geometric construction of non-structurally symmetric strings. To explain what these are consider the Taylor coefficients of a function. These define the components of a scalar string, and considered as a collection of indexed objects they must have symmetric indices as the various partial derivatives of a function are symmetric. It was known, from the work of Barndorff-Nielsen and Blæsild (1987) that is was possible to have strings with the same indexing but without the symmetry requirement. Jupp (1992) gave a geometric construction of these strings using the theory of *semi-holonomic jets.*

9.6 Exercises for Chapter 9

Exercise 9.1 Let Y be a yoke. Show that $\psi_p(q) = d_pY(\ ,q)$ and $\psi_p(q) = d_qY(p,\ ) - d_qY(q,\ )$ are co-ordinate strings. Calculate them for the expected and observed yokes.

Exercise 9.2 Let S^1 be the unit circle in $\mathbf{R}^2$ parametrised by $(\cos(\theta), \sin(\theta)$. Show that the tangent space at $((\cos(\theta), \sin(\theta)$ is spanned by the vector $(-\sin(\theta), \cos(\theta))$. Define a co-ordinate string by projecting $(\cos(\phi), \sin(\phi)) - (\cos(\theta), \sin(\theta))$ onto the tangent space at $(\cos(\theta), \sin(\theta))$. That is

$$\chi_\theta(\phi) = \langle(\cos(\phi), \sin(\phi)) - (\cos(\theta), \sin(\theta)), (-\sin(\theta), \cos(\theta))\rangle$$

Define a co-ordinate string $\chi_p(q)$ to be *k-flat* if there exist co-ordinates x such that $\chi_p(q)$ and $x(p) - x(q)$ agree to order up to and including k at every p. By Taylor expanding $\chi_\theta(\phi)$ in $\theta - \phi$ show that $\chi_\theta(\phi)$ is 2-flat but not 3-flat.

References

Amari, S-I. (1985) *Differential-Geometrical Methods in Statistics.* Lecture Notes in Statistics **28**, Springer-Verlag, Berlin.

Amari, S-I, Barndorff-Nielsen, O.E., Kass, R.E., Lauritzen, S.L. and Rao, C.R. (1987) *Differential Geometry in Statistical Inference.* Institute of Mathematical Statistics, Lecture Notes – Monograph Series, Volume 10, Institute of Mathematical Statistics, Hayward, California.

Barndorff-Nielsen, O.E. (1978) *Information and exponential families in statistical theory.* Wiley, New York.

Barndorff-Nielsen, O.E. (1986a) Likelihood and observed geometries. *Ann. Statist.* **14**, 856–873.

Barndorff-Nielsen, O.E. (1986b) Strings, tensorial combinants, and Bartlett adjustments. *Proc. R. Soc. London A* **406**, 127–137.

Barndorff-Nielsen, O.E. (1987) Differential geometry and statistics: some mathematical aspects. *Indian J. Math.* **29**, Ramanujan Centenary Volume.

Barndorff-Nielsen, O.E. (1986) Strings, tensorial combinants, and Bartlett adjustments. *Proc. R. Soc. London A* **406**, 127-137.

Barndorff-Nielsen, O.E. (1988) *Parametric Statistical Models and Likelihood.* Lecture Notes in Statistics **50**, Springer-Verlag, Berlin.

Barndorff-Nielsen, O.E. and Blæsild, P. (1987) Derivative strings: contravariant aspect. *Proc. R. Soc. London A* **411**, 421–444.

Barndorff-Neilsen, O.E., Blæsild, P, Carey, A.L, Jupp, P.E., Mora, M. and Murray, M.K. (1992) Finite dimensional algebraic representations of the infinite phylon group. *Acta. Appl. Math.* **28**, 219–252.

Barndorff-Nielsen, O.E., Blæsild, P and Eriksen, P.S. (1989) *Decomposition of Measures, and Statistical Transformation Models.* Lecture Notes in Statistics **58**, Springer-Verlag, Berlin.

Barndorff-Nielsen, O.E. and Cox, D.R. (1984) Bartlett adjustments to the likelihood ratio statistic and the distribution of the maximum likelihood estimator. *J. R. Statist. Soc. Ser. B* **46**, 483–495.

Bott, R and Tu L.W. (1982) *Differential forms in algebraic topology.* Graduate Texts in Mathematics **82**, Springer-Verlag, Berlin.

Cox, D.R. and Hinkley, D.V. (1974) *Theoretical Statistics.* Chapman and Hall, London.

Critchley, F., Marriott, P.K. and Salmon, M. (1991). Preferred point geometry and statistical manifolds. Preprint, University of Warwick.

Eells Jr, J. (1966) A setting for global analysis. *Bull. Amer. Math. Soc.* **72**, 751–807.

Efron, B, (1975) Defining the curvature of a statistical problem (with applications to second order efficiency). *Ann. Statist.* **3**, 1189-1242.

Fleming, W.H. (1987) *Functions of Several Variables.* Springer Undergraduate Texts in Mathematics, Springer-Verlag, Berlin.

Foster, B.L. (1958) Differentiation on manifolds without a connection. *Michigan Math. J.* **5**, 183–190.

Foster, B.L. (1989) Higher derivatives in geometry and physics. *Proc. R. Soc. London A* **423**, 443–455.

Golubitsky, M. and Guillemin, V. (1973) *Stable Mappings and their Singularities.* Graduate Texts in Mathematics **14**, Springer-Verlag, Berlin.

Helgason, S. (1978) *Differential Geometry, Lie groups and Symmetric Spaces.* Academic Press, New York.

Humphreys, J.E. (1981) *Linear Algebraic Groups.* Graduate Texts in Mathematics, **21**. Springer-Verlag, Berlin.

Jupp, P.E. (1992) Derivative strings, differential strings and semi-holonomic jets. *Proc. R. Soc. London A* **436**, 89–98.

Kass, R.E. (1984) Canonical parametrisation and zero parameter effects curvature. *J. Roy. Statist. Soc. B* **46**, 86–92.

Klingenberg, W. (1978) *A course in differential geometry.* Graduate Texts in Mathematics **51**, Springer-Verlag, Berlin.

Koboyashi, S. (1961) Frame bundles and higher order contact. *Proc. Symp. Pure Math.* **3**, 186–193, American Mathematical Society, Providence, Rhode Island.

Koboyashi, S, and Nomizu, K. (1963) *Foundations of Differential*

Geometry. Volumes 1 and 2. Interscience, Wiley, New York.

Lang, S. (1967) *Introduction to Differentiable Manifolds.* Interscience, Wiley, New York.

Marriott, P.K. (1987) *Applications of Differential Geometry to Statistics.* Ph.D. thesis, University of Warwick.

McCullagh, P and Cox, D.R. (1986) Invariants and likelihood ratio statistics. *Ann. Statist.* **14**, 1419–1430.

Murray, M.K. (1986) Taylor series and co-ordinate systems in statistics. *Proc. R. Soc. Lond. A* **415**, 445–452.

Rao, C.R. (1945) Information and the accuracy attainable in the estimation of statistical parameters. *Bull. Calcutta Math. Soc.* **37**, 81–91.

Spivak, M. (1970) *A Comprehensive Introduction to Differential Geometry.* Volumes 1–5, Publish or Perish, Boston.

Terng, C.L. (1978) Natural vector bundles and natural differential operators. *Amer. J. Math.* **100**, (4), 775–828.

Vos, P.W. (1989) Fundamental equations for statistical submanifolds with applications to the Bartlett adjustment. *Ann. Inst. Statist. Math.* **41**, No 3, 429–450.

Wald, A. (1949) Note on the consistency of the maximum likelihood estimate. *Ann. Math. Statist.* **20**, 595–601.

Notation index

Chapter 1

$p(\theta)$, 1
$\mathbf{R}^d$, 1
$N(\mu, \sigma^2)$, 1
$\mathcal{N}$, 1
$K(\theta)$, 2
μ, 2
$d\mu$, 2
$x^1, \ldots, x^r$, 2
$\theta^1, \ldots, \theta^r$, 2
Ω, 9
$\mathcal{M}$, 9
ℓ, 10
R_Ω, 10
$\mathcal{P}$, 10
$\mathcal{N}(\mathbf{R}^n, 1)$, 13
$\mathbf{R}.1$, 15
$R_\Omega/\mathbf{R}.1$. 15
θ^0, 15
$\tilde{P}$, 15
$\tilde{\ell}$, 16
$\tilde{\ell}_i$, 16
$\tilde{\ell}_{ij}$, 16
$\langle f, \quad g\rangle_p$, 18
$E_p(fg)$, 18
g_{ij}, 18
N_p, 18
$\pi_p(f)$, 18
Σ, 18
α_{ij}, 19
$\gamma(p)$, 20
$\mathcal{N}(\mathbf{R}^n)$, 24
$p(\mu, \Sigma)$, 24

Chapter 2

ℓ_x, 28
$\mu(p)$, 28
$f \circ \gamma$, 29
$f \circ \gamma'(0)$, 29
$\sigma(p)$, 29
γ_i, 31
$\bar{f}$, 31
$\gamma'(0)$, 37
T_pP, 37
$T_a\mathbf{R}^n$, 38
T_pV, 39
$T_\nu\mathcal{M}$, 41
df, 46

$d\phi^i$, 46

TP, 47

$\pi: TP \to P$, 47

$\partial/\partial\theta^i$, 47

δ^i_j, 49

$\mathbf{i}_r$, 53

$\mathbf{i}_\theta$, 53

$E(f)$, 62

Chapter 3

F_α, 63

$F: M \to N$, 65

T_pF, 67

$SO(n)$, 71

$i: Q \to P$, 74

α, 80

Chapter 4

∇, 100

Γ^i_{jk}, 107

π, 116

$E^\alpha_p(f)$, 120

∇^α, 121

$\Gamma^{\alpha i}{}_{jk}$, 121

$\exp_p$, 129

Chapter 5

T_γ, 136

$\wedge$, 140

d, 143

R, 153

Chapter 6

$g(v, w)$, 161

length, 161

$d(x, y)$, 161

g_{ij}, 162

$[ij : k]$, 168

∇^0, 169

L_g, 179

$SL(2, \mathbf{R})$, 179

$SO(2)$, 181

M, 182

Hess_p, 182

Y, 183

$\sqrt{p}$, 185

δ_i, 186

$\Delta(T_pP)$, 189

$\Delta(P)$, 189

Chapter 7

Ω^N, 194

μ^N, 194

p^N, 194

M_N, 195

η^i, 197

cov, 202

$v_x(\mu)$, 204

$E(X|Y)$, 205

R_Y, 207

D^α_η, 218

$\nabla^{(k)}_M$, 218

$\nabla^{(k)}_E$, 219

f^*, 219

p^*, 221

Chapter 8

T^*P, 226
$\mathcal{H}$, 229
$\otimes$, 230
$\oplus$, 230
$F(V)$, 231
$GL(n)$, 231
$F(P)$, 232
$\mathcal{G}_\infty(n)$, 234
$\mathcal{F}^\infty(P)$, 235
$\mathcal{G}_r(n)$, 235
$\mathcal{F}^r(P)$, 235
$[f, v]$, 236
$T^{(r,s)}$, 239

Chapter 9

$J_p^r(P, \mathbf{R})$, 244
$J^r(P, \mathbf{R})$, 244
I^k, 245
$P^k(\mathbf{R}^n)$, 246
$S^k(V^*)$, 246
$T_\varphi^k(f)$, 250
$\nu_{i,j,k}$, 253
$\mathcal{G}^{(T)}$, 256
$\mathcal{G}_T^{(1)}$, 257
$\mathcal{C}_T$, 257
Hom, 259
$\mathbf{C}P_r$, 262

Subject index

2-frames, 247
*i*th co-ordinate variation, 31
p^* formula, 221
$SO(n)$, 71

affine co-ordinate system, 8, 60
affine spaces, 6
  flat connection on, 109
alpha connection, 121
alpha embedding, 63, 120
alpha expectation, 120
Amari's 1-connection, 117
Amari's α-connection, 120
associated bundles, 235
asymptotically unbiased estimators, 196
asymptotics of mle, 213
  for curved exponential family, 219
  for unit normal family, 214
asymptotics, 194

bijection, 65
binomial family, 11

canonical parameters, 2, 4
central limit theorem, 211
chain rule, 35, 54
Christoffel symbols of a connection, 107
codomain of a function, 65
co-ordinate expression of a function, 33
co-ordinate string, 250
co-ordinates, 2, 26
  affinely related, 8, 60
  affine, 8, 60
  smoothly related, 44
  calculations in, 50

complete exponential family, 197
components of a co-ordinate string, 252
connection 1-forms, 106, 151
connection, 97, 105
  induced on a submanifold, 111
  on the cotangent bundle, 121
  dual pair of, 125
  geodesic for, 126
  symmetric, 123
  metric preserving, 165
  Levi-Civita, 168, 187
  on a vector bundle, 230
Cramer–Rao inequality, 201
curvature and flatness, 154
curvature, 132, 148, 153
curved exponential family, 69, 83
  asymptotics of mle, 219
  mle, 199
  score of, 79
cylindrical connection, 107

densities, 188
derivative map, 67
differential 1-forms, 45
differential 2-forms, 139
differential k-forms, 147
domain of a function, 64
dual connection, 125, 170

embedding, 73
estimators, 195
  asymptotically unbiased, 196
  maximum-likelihood, 182, 196
  unbiased, 195

expectation co-ordinates, 197
exponential families, 2,
  mle, 197
  score, 78
exponential map, 129
exponential transformation
  model, 92

factorisation principle, 206
Fisher information matrix, 18
Fisher information metric, 163
flat connection, 110, 154
flatness of a co-ordinate string,
  252
frame bundle, 231
frame for a vector bundle, 225

geodesic, 126
  on the sphere, 127
graded space, 247
group, 86

half-densities, 191
Hilbert bundle, 229

immersion, 72
imbedding curvature, 19
inclusion map, 74
infinite frame bundle, 234
information metric, 163
injection, 65
integration on a manifold, 188
inverse mapping theorem, 70

jet bundles, 243
jet equivalence, 243
jet of a diffeomorphism, 235
jet, 234, 243

Koszul connection, 105

length of a curve, 161
Levi-Civita connection, 115,
  168, 187
Lie group action, 87
Lie group, 87
likelihood ratio statistic, 208
limit theorems, 211
location scale group, 86
log-likelihood, 10

manifold, 57
matrix group, 87
maximum-likelihood estimator
  (mle), 182, 196
metric preserving connections,
  165
minimal sufficiency, 208
mixture affine structure, 200
mle (maximum likelihood
  estimator), 182, 196

natural co-ordinate string, 251
normal bundle, 229
normal family, 1, 17
  mle, 198

observed geometry, 181
orbit of a group action, 90

parallel transport, 135
phylon group, 234
phylon representations, 255
Poincáre lemma, 147
polynomials on a vector space,
  246
principal bundle, 231
product manifold, 84
proper embedding, 73
pullback, 65, 145
pushout of measure, 204

representation theory of $GL(n)$,
  240
representations, 89
Riemannian manifold, 161
Riemannian metric, 161

Schur's lemma, 261
score, 77
second fundamental form, 19,
  80, 116, 119, 130
second jet bundle, 244
skewness tensor, 241
smooth map, 64
square root likelihood, 185
  and half-densities, 191
  and Levi-Civita connection,
  187
statistical curvature, 20
statistical divergences, 173
  minimisation, 176
statistical manifold, 76

strings, 254
submanifold, 68
submersion theorem, 70
sufficiency, 204
sufficient statistics, 206
surjection, 65
symmetric connections, 123

tangent bundle, 47, 223
tangent space, 37
tangent vector, 37
  in an affine space, 40
  in a vector space, 39
  to a parametrised family, 42
  to probability distributions, 42
  in $\mathbf{R}^n$, 38
Taylor series, 248
tensors, 239
transformation model, 91, 179
transitive group action, 90

unbiased estimators, 196
unit normal family, 13
  asymptotics of mle, 213

variation, 29
vector bundles, 226
  frame for, 225
  local section of, 225
  section of, 225
vector field, 48
vector group, 86
velocity, 35
von Mises–Fisher model, 91

weak law of large numbers, 211

yokes, 184